T0222744

Introduction to Prescriptive AI

A Primer for Decision Intelligence Solutioning with Python

Akshay Kulkarni
Adarsha Shivananda
Avinash Manure

Apress®

Introduction to Prescriptive AI: A Primer for Decision Intelligence Solutioning with Python

Akshay Kulkarni
Bangalore, Karnataka, India

Adarsha Shivananda
Hosanagara, Karnataka, India

Avinash Manure
Bangalore, Karnataka, India

ISBN-13 (pbk): 978-1-4842-9567-0
https://doi.org/10.1007/978-1-4842-9568-7

ISBN-13 (electronic): 978-1-4842-9568-7

Managing Director, Apress Media LLC: Welmoed Spahr
Acquisitions Editor: Celestin Suresh John
Development Editor: Laura Berendson
Coordinating Editor: Mark Powers

Cover designed by eStudioCalamar

Cover image by Anna Tsukanova on Unsplash (www.unsplash.com)

Distributed to the book trade worldwide by Apress Media, LLC, 1 New York Plaza, New York, NY 10004, U.S.A. Phone 1-800-SPRINGER, fax (201) 348-4505, e-mail orders-ny@springer-sbm.com, or visit www.springeronline.com. Apress Media, LLC is a California LLC and the sole member (owner) is Springer Science + Business Media Finance Inc (SSBM Finance Inc). SSBM Finance Inc is a **Delaware** corporation.

For information on translations, please e-mail booktranslations@springernature.com; for reprint, paperback, or audio rights, please e-mail bookpermissions@springernature.com.

Apress titles may be purchased in bulk for academic, corporate, or promotional use. eBook versions and licenses are also available for most titles. For more information, reference our Print and eBook Bulk Sales web page at www.apress.com/bulk-sales.

Any source code or other supplementary material referenced by the author in this book is available to readers on GitHub (https://github.com/Apress). For more detailed information, please visit www.apress.com/source-code.

Printed on acid-free paper

This book is dedicated to my 2-year-old daughter, Siri, who has been a constant source of inspiration, joy, and wonder in my life since the day she was born. Siri, you have shown me the true meaning of unconditional love, and watching you grow and learn has been the greatest adventure of my life. This book is a tribute to the amazing person you are becoming and a reminder of the infinite potential that lies within you. Thank you for being my daughter, my teacher, and my greatest blessing.

—Avinash Manure

Table of Contents

About the Authors

Akshay Kulkarni is an artificial intelligence (AI) and machine learning (ML) evangelist and a thought leader. As a consultant, he has worked with several Fortune 500 and global enterprises to drive AI and data science–led strategic transformations. He is a Google developer, an author, and a regular speaker at major AI and data science conferences (including the O'Reilly Strata Data & AI Conference and Great International Developer Summit [GIDS]). He is a visiting faculty member at some of the top graduate institutes in India. In 2019, he was featured as one of India's "top 40 under 40" data scientists. In his spare time, Akshay enjoys reading, writing, coding, and helping aspiring data scientists.

Adarsha Shivananda is a data science and MLOps leader. He is working on creating world-class MLOps capabilities to ensure continuous value delivery from AI. He aims to build a pool of exceptional data scientists within and outside organizations to solve problems through training programs. He always wants to stay ahead of the curve. Adarsha has worked extensively in the pharma, healthcare, CPG, retail, and marketing domains. He lives in Bangalore and loves to read and teach data science.

Avinash Manure is a seasoned machine learning professional with 10+ years of experience building, deploying, and maintaining state-of-the-art machine learning solutions across different industries. He has six years of experience leading and mentoring high-performance teams in developing ML systems catering to different business requirements. He is proficient in deploying complex machine learning and statistical modeling algorithms/techniques for identifying patterns and extracting valuable insights for key stakeholders and organizational leadership.

About the Technical Reviewer

Nitin Ranjan Sharma is a manager at Novartis. He leads a team that develops products using multimodal techniques. As a consultant, he has developed solutions for Fortune 500 companies and has been involved in solving complex business problems using machine learning and deep learning frameworks. His major focus area and core expertise is computer vision, including solving challenging business problems dealing with images and video data. Before Novartis, he was part of the data science team at Publicis Sapient, EY, and TekSystems Global Services. He is a regular speaker at data science community meetups and an open-source contributor. He also enjoys training and mentoring data science enthusiasts.

Acknowledgments

We would like to thank our families who have always made sure we had the right environment at home to concentrate on this book and complete it on time. We would also like to thank the publishing team—Mark Powers and Celestin Suresh John, and our technical reviewer, Nitin Sharma—who helped us make sure this book was the best it could be. We would also like to thank our mentors who made sure we grew professionally and personally by always supporting us in our dreams and guiding us toward achieving our goals. Last but not least, we thank our parents, our friends, and our colleagues who were always there in tough times and motivated us to chase our dreams.

Introduction

This book will introduce you to the concept of decision intelligence, including its history and current and future trends. It will help you evaluate different decision intelligence techniques and guide you on how to them through different prescriptive AI methods and incorporate them into business workflows through different domain-specific use case implementations.

This book is for data scientists, AI/machine learning engineers, and deep learning professionals who are working toward building advanced intelligent AI/ML applications. This book is also for business professionals and nontechnical stakeholders who want to understand how decision intelligence can help a business grow.

This book will take you through the journey of decision-making in companies with key milestones, key statistics, and benefits. It will provide insights on where decision intelligence fits within the AI life cycle. This book will provide insights on how to prepare for prescriptive AI (a key requirement to decision intelligence) with the help of a business requirement document. It will then deep dive into different decision intelligence methodologies, their advantages, and their limitations. Next, you will learn how to perform different simulations and interpret the results from them. Then you will be guided on how to enable and embed the decision intelligence process into the business workflow through prescriptive AI. You will learn about different cognitive biases that humans make and how that can be lowered/eliminated through the combination of machine and human intelligence. Finally, you will find different cases studies by domain through tailored use cases.

By the end of the book, you will have a solid understanding of decision intelligence and prescriptive AI tools and techniques and how to incorporate them within the business workflow for greater productivity and profit to the business.

The source code for this book is available on GitHub (`github.com/apress/introduction-prescriptive-ai`).

CHAPTER 1

Decision Intelligence Overview

Prescriptive AI is a type of artificial intelligence that is designed to provide recommendations, solutions, or actions to optimize or improve a specific process or outcome. It is unlike descriptive AI, which describes "what" has happened in the past/present; inferential/diagnostic AI, which helps us understand "why" something has happened; and predictive AI, which helps predict what might happen in the future. Once we have the predictions, prescriptive AI focuses on what actions should be taken to achieve a particular goal or outcome.

Prescriptive AI can be used in a variety of applications, such as in healthcare to help doctors diagnose diseases and prescribe treatments, in finance to make investment decisions, and in manufacturing to optimize production processes. It typically uses machine learning algorithms and other advanced technologies to analyze large datasets and generate recommendations based on the data.

Overall, the goal of prescriptive AI is to help humans make better decisions by providing them with accurate and actionable insights based on data-driven analysis.

© Akshay Kulkarni, Adarsha Shivananda, Avinash Manure 2023
A. Kulkarni et al., *Introduction to Prescriptive AI*,
https://doi.org/10.1007/978-1-4842-9568-7_1

As mentioned, prescriptive AI systems differ from descriptive and predictive AI systems in that they not only provide insights and predictions based on data analysis but also offer specific recommendations or actions. Prescriptive AI goes a step further by suggesting optimal solutions and strategies to achieve desired outcomes.

In this chapter, we will start with the types of AI and then deep dive into prescriptive AI and decision intelligence.

Types of AI

Descriptive, diagnostic, predictive, and prescriptive AI are four different types of artificial intelligence that are used in various applications. Here's an overview of each:

- **Descriptive AI:** Descriptive AI, also known as *descriptive analytics*, is used to describe what has happened in the past or is currently happening. It involves analyzing historical data to identify patterns, trends, and insights. Descriptive AI is often used for reporting and business intelligence applications, such as dashboards and scorecards that summarize key performance metrics. It can also be used for exploratory data analysis to uncover hidden insights and correlations.

  For example, a retail company may use descriptive AI to analyze its sales data and identify which products are selling the most, which stores are performing the best, and which customers are buying the most. This information can then be used to make data-driven decisions on inventory management, pricing strategies, and marketing campaigns.

- **Diagnostic AI:** Diagnostic AI, also known as *diagnostic analytics*, is used to identify the root cause of a problem or issue. It involves analyzing data to identify factors that contribute to a specific outcome or event. Diagnostic AI is often used in healthcare, finance, and manufacturing applications.

  For example, a hospital may use diagnostic AI to analyze patient data and identify the factors that contribute to re-admissions. This could involve analyzing patient demographics, medical histories, and treatment plans to identify common risk factors and develop targeted interventions to reduce re-admissions.

- **Predictive AI:** Predictive AI, also known as *predictive analytics*, is used to predict what is likely to happen in the future. It involves using statistical modeling and machine learning techniques to analyze historical data and make predictions about future outcomes. Predictive AI is often used in marketing, finance, and customer service applications.

  For example, a bank may use predictive AI to analyze customer data and predict which customers are most likely to default on their loans. This information can then be used to proactively reach out to those customers and offer them personalized loan modification options to avoid default.

- **Prescriptive AI:** Prescriptive AI, also known as *prescriptive analytics*, is used to provide recommendations on what actions to take based on the predictions made by predictive AI. It involves

3

using optimization algorithms to identify the best course of action based on a set of constraints and objectives. Prescriptive AI is often used in supply chain management, logistics, and resource-planning applications.

For example, a logistics company may use prescriptive AI to optimize its delivery routes based on real-time traffic data, weather conditions, and delivery deadlines. This information can then be used to automatically reroute delivery trucks to avoid traffic jams and ensure on-time delivery.

In summary, as shown in Figure 1-1, descriptive AI is used to describe what has happened, diagnostic AI is used to identify the root cause of a problem, predictive AI is used to predict what is likely to happen, and prescriptive AI is used to recommend what actions to take based on those predictions. Each type of AI has different applications and can be used to solve different types of problems.

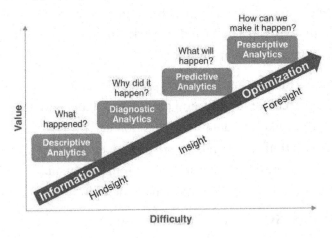

Figure 1-1. Types of AI

Prescriptive AI typically involves analyzing large amounts of data using machine learning algorithms to identify patterns and relationships between variables. It then uses this analysis to provide recommendations or decisions that can optimize business processes, improve customer experiences, or achieve other desired outcomes.

Examples of prescriptive AI applications include personalized medicine, where AI systems provide doctors with recommendations for personalized treatments based on a patient's genetic and medical history, and supply chain management, where AI systems optimize logistics and inventory management to minimize costs and improve efficiency.

We try to achieve preceptive AI through decision intelligence.

Decision Intelligence

Decision intelligence (DI) is a discipline that combines techniques from artificial intelligence, decision theory, and behavioral science to support and improve human decision-making. It aims to provide a more structured and rigorous approach to decision-making by using data and analytics to help people make better decisions. It involves using a combination of human expertise and machine intelligence to make better decisions in complex and uncertain situations.

DI involves a variety of techniques, including machine learning, optimization, simulation, and behavioral science, to provide decision-makers with insights and recommendations. It typically involves collecting and analyzing data from various sources, identifying decision criteria, and generating possible alternatives based on various constraints and objectives.

The ultimate goal of DI is to enable organizations and individuals to make better decisions that are more aligned with their goals, values, and preferences. This can lead to improved efficiency, effectiveness, and competitiveness in a wide range of domains, including business, healthcare, government, and education.

5

Decision Intelligence History

Decision intelligence has its roots in a variety of fields, including decision theory, game theory, and operations research. Let us understand a brief history of DI.

- **Decision theory:** Decision theory is a field of study that has been around for more than a century. In the early 1900s, economists and mathematicians began to develop formal models for decision-making, which included elements such as utility theory and expected value. These models provided a framework for understanding how decisions are made and how outcomes can be optimized.

- **Operations research:** Operations research (OR) emerged during World War II as a way to solve military logistics problems. OR involves using mathematical and statistical methods to analyze complex systems and make decisions. OR has been applied to a wide range of industries, including manufacturing, transportation, and healthcare.

- **Game theory:** Game theory is a branch of mathematics that studies how people interact in strategic situations. It was developed in the mid-20th century by mathematicians such as John Nash and John von Neumann and has been used to study decision-making in fields such as economics, political science, and psychology.

- **Business analytics:** In recent years, with the rise of big data and data analytics, decision-making has become increasingly data-driven. Business analytics involves using data to inform decision-making and has been used to optimize processes, improve customer experiences, and drive revenue growth.

- **Decision intelligence:** DI emerged in the early 2000s as a way to integrate the various disciplines involved in decision-making, including decision theory, game theory, and analytics. DI involves using a range of tools and techniques to support decision-making, including machine learning, optimization, and simulation.

Challenges in AI Adoption

There are several challenges that businesses face when adopting AI technology. These are some of the key challenges:

- **Data quality and availability:** AI systems rely heavily on data, so poor data quality and a lack of available data can hinder the effectiveness of the system.

- **Technical expertise:** Implementing/maintaining AI systems often requires specialized technical expertise, which may not be readily available within the business.

- **Cost:** AI systems can be expensive to implement and maintain, especially if the business does not have the necessary infrastructure in place.

- **Resistance to change:** There may be resistance to change from employees who are used to working with traditional systems and processes.

- **Ethical considerations:** AI systems can raise ethical concerns related to privacy, bias, and transparency, which must be addressed to ensure ethical decision-making.

- **Integration with existing systems:** Integrating AI with existing systems can be challenging, particularly if those systems are outdated or incompatible with the new technology.

- **Regulation and compliance:** AI systems may need to comply with regulations such as GDPR or HIPAA, which can add an extra layer of complexity to their implementation.

- **Lack of understanding:** There may be a lack of understanding among business leaders and employees about the potential benefits and limitations of AI technology, which can hinder adoption.

To overcome these challenges, businesses need to carefully evaluate the potential benefits and risks of AI adoption, identify the necessary resources and expertise required, and develop a comprehensive plan for implementing and integrating the technology into their existing systems and processes. They also need to prioritize ethical considerations and work closely with employees to ensure that they are comfortable and confident using the new technology.

How Can DI Help Bridge the Gap Between AI and Business?

One of the challenges with AI is that it can be difficult to understand how it arrives at its decisions. DI can help by providing insights into how AI models work, allowing business users to understand and interpret the decisions that are being made.

- **Integration:** DI can help integrate AI technology into existing business processes and workflows, making it easier for business users to take advantage of AI capabilities without having to learn complex new systems.

- **Customization:** DI can be customized to fit the specific needs of a business, allowing for a more personalized approach to decision-making that takes into account the unique challenges and opportunities facing the organization.

- **Risk management:** DI can help identify and mitigate risks associated with AI technology, such as bias, privacy concerns, and data quality issues, ensuring that the benefits of AI are realized while minimizing the risks.

- **Transparency:** DI can help improve transparency in decision-making by providing clear and understandable explanations of the decisions being made, helping to build trust and confidence in AI technology among business users and stakeholders.

By leveraging the power of AI in a way that is tailored to the specific needs of businesses, DI can help bridge the gap between AI and business, enabling organizations to take advantage of the benefits of AI technology while minimizing the risks and ensuring that decisions are made in a way that is transparent, ethical, and effective.

The Need for Decision Intelligence

There are several reasons why DI is becoming increasingly important in today's world.

- **Information overload:** With the explosion of data and information, decision-makers are often overwhelmed with too much information, making it difficult to make informed decisions. DI can help cut through the noise by analyzing and synthesizing relevant data into actionable insights.

- **Complex and uncertain environments:** Many decisions today are made in complex and uncertain environments, where there are multiple factors and variables to consider. DI can help decision-makers better understand the interdependencies and trade-offs between different options and provide recommendations based on probabilistic modeling and simulation.

- **Cognitive biases:** Human decision-making is often subject to cognitive biases, such as overconfidence, anchoring, and confirmation bias. DI can help mitigate these biases by providing a more objective and data-driven approach to decision-making.

- **Competitive advantage:** In today's competitive environment, organizations that can make better decisions faster are more likely to succeed. DI can provide a competitive advantage by enabling organizations to make better decisions based on data-driven insights.

- **Improved decision quality:** DI can help decision-makers make better, more informed decisions by providing insights and recommendations based on data and analytics. This can lead to improved outcomes and increased efficiency.

- **Risk management:** Many decisions involve some level of risk. DI can help decision-makers understand and manage this risk by providing probabilistic assessments and scenario analysis.

- **Agility and flexibility:** In today's fast-paced and rapidly changing environment, organizations need to be agile and flexible in their decision-making. DI can help organizations respond quickly and effectively to changing circumstances by providing real-time data and insights.

- **Transparency and accountability:** DI can help ensure that decisions are transparent and accountable by providing a clear audit trail and documentation of the decision-making process.

- **Personalization:** DI can help personalize decisions based on individual preferences and goals, leading to a more tailored and effective decision-making process.

Overall, DI can help organizations and individuals make better decisions that are more aligned with their goals, values, and preferences, leading to improved outcomes and performance.

The Evolution of Decision-Making

Companies have been shaped by various factors, including advancements in technology, changes in organizational structures, and the growing importance of data and analytics. Here are some key milestones in the evolution of decision-making in companies:

- **Intuition-based decision-making:** In the early days of business, decision-making was often based on the intuition and experience of leaders and managers. Decisions were made quickly and informally, based on a hunch or gut feeling.

- **Rule-based decision-making:** As companies grew and became more complex, decision-making became more structured and rule-based. Managers developed standardized processes and procedures for decision-making, based on predetermined criteria and rules.

- **Data-driven decision-making:** With the rise of information technology and data analytics, decision-making in companies shifted toward a more data-driven approach. Managers began to use data and metrics to inform their decisions, leveraging tools such as business intelligence, data visualization, and predictive analytics.

- **Collaborative decision-making:** In recent years, decision-making in companies has become more collaborative, with a focus on involving stakeholders from across the organization in the decision-making process. This approach helps to ensure that decisions are made with input from diverse perspectives and that all relevant information and expertise are taken into account.

- **AI-powered decision-making:** As artificial intelligence (AI) and machine learning technologies continue to advance, companies are beginning to explore how they can be used to support decision-making. AI-powered decision-making systems can help companies process large amounts of data quickly and identify patterns and insights that may not be immediately apparent to humans.

Overall, the evolution of decision-making in companies has been marked by a shift toward a more data-driven, collaborative, and technology-enabled approach, with a growing emphasis on using data and analytics to support decision-making processes.

Challenges

Decision intelligence can be challenging to implement because it involves several complex and interconnected processes. Here are a few reasons why DI is hard:

- **Complexity of data:** In many cases, the data that is required for DI is complex, voluminous, and unstructured, making it difficult to collect, clean, and process.

- **Interpretability of AI models:** AI models can be complex and difficult to interpret, which can make it challenging to understand how they arrive at their decisions.

- **Expertise required:** Implementing DI requires specialized expertise in areas such as data science, machine learning, and decision-making theory. Finding and hiring the right talent can be challenging, especially given the high demand for these skills.

- **Integration with existing systems:** DI often requires integration with existing systems, which can be complex and time-consuming, especially if those systems are outdated or incompatible with new technology.

- **Change management:** Implementing DI can require significant changes to existing business processes and workflows, which can be difficult to manage and communicate to employees.

Despite these challenges, DI can provide significant benefits to businesses by enabling more informed and effective decision-making. By carefully planning and managing the implementation process, businesses can overcome these challenges and realize the full potential of DI.

Applications

There are numerous applications of DI across different domains. Here are some examples:

- **Business:** DI can help businesses make better decisions in areas such as pricing, product development, supply chain management, and customer service. For example, DI can help retailers optimize pricing strategies based on customer behavior and competitor analysis.

- **Healthcare:** DI can be used to help healthcare professionals make more informed decisions about patient care, diagnosis, and treatment. For example, DI can help doctors determine the best treatment plan for a patient based on their medical history and symptoms.

- **Government:** DI can be used by government agencies to improve decision-making in areas such as policy development, resource allocation, and emergency response. For example, DI can help emergency responders make faster and more effective decisions during a natural disaster.

- **Finance:** DI can help financial institutions make better decisions in areas such as investment management, risk management, and fraud detection. For example, DI can help investment managers optimize portfolio allocation based on market trends and risk profiles.

- **Education:** DI can be used to improve decision-making in areas such as student performance evaluation, curriculum design, and resource allocation. For example, DI can help teachers personalize learning for students based on their individual needs and learning styles.

- **Marketing:** DI can help marketers make better decisions in areas such as customer segmentation, campaign optimization, and product positioning. For example, DI can help marketers identify the most effective marketing channels for different customer segments.

- **Manufacturing:** DI can be used to optimize manufacturing processes, improve quality control, and reduce waste. For example, DI can help manufacturers identify the most efficient production methods based on variables such as raw material cost, energy consumption, and product quality.

- **Transportation:** DI can be used to optimize transportation systems, improve route planning, and reduce congestion. For example, DI can help transportation planners determine the optimal bus routes based on passenger demand and traffic patterns.

- **Agriculture:** DI can be used to improve crop yields, reduce water usage, and optimize fertilizer application. For example, DI can help farmers determine the optimal planting and harvesting times based on weather patterns and soil conditions.

- **Energy:** DI can be used to optimize energy consumption, reduce emissions, and improve energy efficiency. For example, DI can help energy companies determine the most cost-effective sources of renewable energy based on factors such as location and weather patterns.

Overall, DI has the potential to improve decision-making across a wide range of domains, leading to better outcomes and increased efficiency.

Understanding Where Decision Intelligence Fits Within the AI Life Cycle

Decision intelligence plays a critical role in the AI life cycle, particularly in the later stages of the process. Here's how DI fits within the AI life cycle:

- **Data collection:** The first stage of the AI life cycle involves collecting and cleaning data. This stage is important for decision intelligence because the quality of the data collected will influence the quality of the decisions made.

- **Data preparation:** Once data is collected, it must be processed and prepared for use in AI models. This stage may involve data transformation, feature engineering, and data augmentation. These steps can impact the accuracy and effectiveness of decision intelligence.

- **Model training:** This stage involves training AI models using machine learning algorithms. Decision intelligence is critical at this stage, as it involves selecting the right algorithms, defining the problem and objectives, and validating the accuracy and reliability of the models.

- **Model deployment:** Once models are trained, they need to be deployed in production systems. Decision intelligence is important at this stage to ensure that the models are deployed in a way that maximizes their impact and effectiveness while minimizing risk.

- **Monitoring and optimization:** AI models must be continuously monitored and optimized to ensure that they are performing as intended. Decision intelligence plays a critical role in this stage, as it involves analyzing performance metrics, identifying areas for improvement, and making decisions to optimize the models.

- **Decision-making:** Finally, decision intelligence is used to inform decision-making based on the insights and predictions generated by the AI models. This may involve selecting the best course of action based on the predictions generated by the models or using the models to inform strategic decision-making processes.

Overall, decision intelligence is a critical component of the AI life cycle, as it helps to ensure that AI models are developed, deployed, and used in a way that maximizes their impact and effectiveness.

Decision Intelligence Methodologies

There are several methodologies that can be used in decision intelligence, depending on the specific problem being addressed and the data available. Here are some commonly used methodologies:

- **Decision analysis:** This methodology involves structuring a decision problem into a decision tree, identifying possible alternatives, evaluating the outcomes of each alternative, and assigning probabilities to each outcome. This helps decision-makers understand the trade-offs between different options and choose the best course of action.

- **Machine learning:** This methodology involves using algorithms to learn patterns in data and make predictions or decisions based on those patterns. Machine learning can be used for tasks such as predictive modeling, classification, and clustering.

- **Optimization:** This methodology involves finding the best solution to a problem that maximizes or minimizes a particular objective function, subject to certain constraints. Optimization can be used for tasks such as resource allocation, production planning, and scheduling.

- **Simulation:** This methodology involves creating a computer model of a system or process and using it to generate data and test different scenarios. Simulation can be used for tasks such as risk assessment, performance evaluation, and process improvement.

- **Game theory:** This methodology involves modeling decision-making in competitive situations, such as auctions, negotiations, and pricing. Game theory can be used to predict the behavior of other players and choose the best strategy based on that prediction.

- **Bayesian inference:** This methodology involves updating prior beliefs based on new data to make probabilistic predictions or decisions. Bayesian inference can be used for tasks such as fraud detection, customer segmentation, and anomaly detection.

- **Multicriteria decision analysis (MCDA):** This methodology involves evaluating alternatives based on multiple criteria or attributes and weighting each criterion based on its importance. MCDA can be used to support complex decisions involving multiple stakeholders with different preferences.

- **Fuzzy logic:** This methodology involves dealing with imprecise or uncertain data by using fuzzy sets and fuzzy rules. Fuzzy logic can be used to model decision-making in situations where data is incomplete or vague, such as in natural language processing.

- **Expert systems:** This methodology involves encoding human expertise and knowledge into a computer system to provide decision support. Expert systems can be used to diagnose problems, recommend solutions, and provide explanations for their reasoning.

- **Heuristics:** This methodology involves using rules of thumb or shortcuts to make decisions quickly and efficiently, based on past experience or common sense. Heuristics can be used to support decisions in situations where time or resources are limited.

These methodologies can be combined and customized to fit specific decision problems and datasets, providing a more rigorous and structured approach to decision-making in complex and uncertain environments.

Let us understand how DI systems are used in a real-world scenario once a model has made its predictions.

Suppose the financial institution from the previous example has used DI to develop a predictive model to identify which customers are most likely to use a new credit card product. The model has predicted that customers who frequently shop at certain retailers are the most likely to use the new credit card.

- **Develop a targeted marketing campaign:** Using the predictions from the model, the financial institution would develop a targeted marketing campaign to promote the new credit card to customers who frequently shop at the identified retailers. This campaign might include personalized offers, targeted advertisements, and incentives to encourage customers to sign up for the new credit card. The financial institution could use data analytics tools to segment the identified customers based on their shopping behavior, demographics, and credit history to tailor the campaign messaging and offers to their specific needs.

- **Monitor the effectiveness of the campaign:** The financial institution would monitor the effectiveness of the marketing campaign by collecting and analyzing customer data such as response rates, conversion rates, and revenue generated by the new credit card product. By using this data, the institution can refine the campaign messaging, improve the targeting of the campaign, and optimize its decision-making processes.

- **Continuously update the model:** As new data becomes available, the financial institution would continuously update the predictive model to improve its accuracy and effectiveness. This might involve incorporating new data sources such as customer feedback, social media data, or external economic indicators. The institution might also adjust the model's algorithms, adjust the weighting of different variables, or use ensemble modeling techniques to improve its performance.

- **Evaluate the impact on the business:** The financial institution would evaluate the impact of the new credit card product on the business by collecting and analyzing data such as revenue generated, customer satisfaction, and retention rates. By using this data, the institution can adjust its strategy, refine its models, and optimize its decision-making processes to maximize its business performance.

By using DI to develop a targeted marketing campaign based on the predictions of a predictive model, the financial institution can maximize the effectiveness of its marketing efforts and increase the chances of success for the new credit card product. By continuously updating the model and evaluating its impact on the business, the institution can further optimize its decision-making processes and improve its overall performance.

Some Potential Pros and Cons of DI

Here are some pros of DI:

- **Improved decision-making:** DI can help businesses make better decisions by providing them with data-driven insights that would be difficult to obtain using traditional decision-making methods.

- **Increased efficiency:** DI can automate certain decision-making processes, which can save businesses time and money.

- **Enhanced accuracy:** DI algorithms can analyze large amounts of data and identify patterns that humans may not be able to see, which can lead to more accurate predictions and better decision-making.

- **Customization:** DI can be tailored to fit the specific needs of a business, allowing for a more personalized approach to decision-making.

- **Improved risk management:** DI can help businesses identify and mitigate risks by analyzing data and predicting potential outcomes.

- **Scalability:** DI can scale to handle large amounts of data and decision-making tasks, which can be difficult or impossible to do manually.

- **Real-time decision-making:** DI can provide real-time insights and decision-making, which can be critical in time-sensitive situations.

- **Competitive advantage:** DI can provide businesses with a competitive advantage by enabling them to make better decisions faster than their competitors.

- **Continuous improvement:** DI algorithms can continuously learn and improve over time, allowing businesses to refine their decision-making processes and stay ahead of the competition.

- **Cost-effective:** In the long run, DI can be cost-effective compared to traditional decision-making methods, especially when dealing with large amounts of data.

Here are some cons of DI:

- **Data quality issues:** DI relies heavily on the quality of the data it uses, and if the data is inaccurate or incomplete, it can lead to flawed decisions.

- **Privacy concerns:** DI often involves collecting and analyzing large amounts of data, which can raise privacy concerns among customers and stakeholders.

- **Bias:** DI algorithms can be biased if they are trained on biased data or if the algorithm itself is biased.

- **Overreliance on technology:** DI can lead to overreliance on technology, which can result in a lack of human input and oversight.

- **Complexity:** DI can be complex and difficult to understand, which can make it challenging for businesses to implement and use effectively.

- **Technical expertise:** Implementing DI requires technical expertise and resources that may be lacking in some businesses.

- **Ethical concerns:** DI raises ethical concerns related to privacy, bias, and transparency, which businesses must address to ensure ethical decision-making.

- **Dependence on data:** DI is dependent on having enough data to make informed decisions, which may not always be available.

- **Integration challenges:** Integrating DI into existing decision-making processes can be challenging and require significant effort.

- **Adoption barriers:** DI adoption may face barriers due to resistance to change or lack of understanding of its potential benefits.

Examples of How Companies Are Leveraging DI

It is difficult to estimate the number of companies that have successfully implemented DI, as the adoption of DI is still relatively new and varies by industry and organization. However, there are several examples of companies that have successfully implemented DI and are seeing significant benefits. Here are a few examples:

- **Capital One:** Capital One, a financial services company, has implemented DI to improve its credit decision-making process. By combining machine learning models with human expertise, Capital One has been able to increase the accuracy of credit decisions and reduce the risk of default.

- **Boeing:** Boeing, an aerospace company, has implemented DI to optimize its aircraft maintenance process. By analyzing data from sensors on aircraft engines, Boeing can predict when maintenance is required and proactively schedule repairs, reducing downtime and improving safety.

- **Procter & Gamble:** Procter & Gamble, a consumer goods company, has implemented DI to improve its supply chain operations. By analyzing data on inventory, production, and logistics, Procter & Gamble can optimize its supply chain to reduce costs and improve delivery times.

- **UPS:** UPS, a logistics company, has implemented DI to optimize its package delivery process. By using machine learning models to predict package demand and optimize delivery routes, UPS has been able to reduce fuel consumption, lower costs, and improve delivery times.

While these are just a few examples, they demonstrate the potential of DI to improve decision-making and drive business value. As the adoption of DI continues to grow, we can expect to see more companies implementing DI and realizing its benefits.

Conclusion

You now understand what prescriptive AI is, its history, how it works, its methodologies, its applications, and its pros and cons. In the coming chapter, we will dig deeper into each of these aspects along with building an end-to-end DI solution.

CHAPTER 2

Decision Intelligence Requirements

Lots of AI projects will fail across industries because AI projects can be complex and challenging, and many potential factors can contribute to their failure. Some of the reasons for failure are lack of clear goals, data quality issues, and difficulty integrating the AI system into existing processes. Additionally, there can be organizational challenges such as a lack of support or buy-in from stakeholders, unrealistic expectations, and a lack of understanding of the capabilities and limitations of AI.

In this chapter, we are going to discuss how we might reduce the failures by planning, having clear requirements, and understanding the DI requirements framework.

Why Do AI Projects Fail?

AI projects can fail for a variety of reasons, including lack of clearly defined goals, lack of data or poor-quality data, lack of expertise or resources, and difficulty in integrating the AI system into existing processes or systems. Additionally, ethical considerations, such as bias in the data or the algorithms themselves, can contribute to the failure of an AI project.

© Akshay Kulkarni, Adarsha Shivananda, Avinash Manure 2023
A. Kulkarni et al., *Introduction to Prescriptive AI*,
https://doi.org/10.1007/978-1-4842-9568-7_2

Decision intelligence (DI) is a field that combines artificial intelligence (AI) and decision-making techniques to help organizations make better decisions. Decision intelligence requirements refer to the specific needs and criteria that must be met for a DI system to be effective and valuable to an organization.

Let's take a step back and understand some of the questions we need to ask before we start any AI projects.

- "Can AI perform better than the current processes?"

- "Is the data good for the AI models?"

- "Are the goals/objectives clearly defined for an AI project?"

- "Is this AI project creating any value?"

- "Did we select the right key performance indicators to measure the value of the AI projects?"

- "Are AI results being consumed (through taking action)?"

- "Are business users getting benefits from AI predictions?"

These are some of the myths surrounding AI:

- **Myth 1:** The right data scientists and the right algorithms create the perfect AI.

- **Myth 2:** To win at AI, you need a lot of data, and you need the best algorithms.

- **Myth 3:** The more accurate the AI model, the better it is.

- **Myth 4:** AI pilots work in one region/domain, and the same AI models work in enterprise rollout.

- **Myth 5:** If the AI model works today, it works forever.

In answer to the previous questions and to overcome these myths, a DI requirements framework is essential. A framework acts as a bible to AI projects because it contains every detail about the project, from the ideation phase to the operationalize phase.

To deliver value through AI, it's important to follow some of the guidelines/paths that can help ensure the successful implementation and adoption of decision intelligence solutions.

- **Clearly define the problem:** Before implementing any AI and decision intelligence solution, it's important to have a clear understanding of the problem that needs to be solved. This means identifying the stakeholders, their needs and requirements, the scope of the problem, and the expected outcomes.

- **Gather and analyze data:** AI solutions rely heavily on data. Therefore, it's important to collect and analyze the required data to ensure that the solution is based on accurate and reliable information.

- **Identify decision-making criteria:** AI should be designed to support decision-making by providing criteria for evaluating alternatives. These criteria should be based on the problem definition, stakeholder requirements, and relevant data.

- **Choose the appropriate decision-making method:** Different decision-making methods can be used depending on the problem and the available data. It's important to choose the appropriate method that will provide the best outcome based on the available information. This will be discussed in detail in Chapter 3.

- **Build the AI model:** Once the criteria and method have been identified, a model can be built. The model should be transparent, interpretable, and user-friendly.

- **Validate the model:** The AI model should be validated to ensure that it's accurate and reliable. This can be done through testing, simulation, and real-time analysis.

- **Implement the solution:** After the model has been validated, the solution can be implemented. It's important to ensure that the solution is integrated with existing systems and that stakeholders are trained and supported to use the solution effectively.

- **Monitor and evaluate the solution:** AI solutions should be continuously monitored and evaluated to ensure that they are delivering the expected value. This can be done through regular performance checks and feedback from stakeholders.

Keeping all these guidelines in mind, let's build a *DI requirements framework* that will act as a handbook to build requirement documents before even starting a project.

DI Requirements Framework

To make AI projects successful, building fancy machine learning and deep learning models and deploying them won't do the trick. That's just the tip of the iceberg. The real effort needs to be put into planning, defining clear objectives, building the approach framework, and evaluating the potential value that could be generated from these projects, as shown in Figure 2-1.

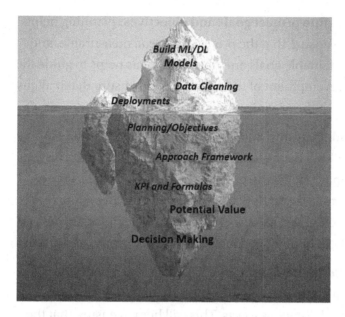

Figure 2-1. *Hidden steps in the AI life cycle*

So, let's discuss a framework in detail. The first step is proper planning.

Planning

Planning is a crucial aspect of any AI project, as it helps to ensure that the project is well-defined, is well-organized, and has a clear path to success. These are some of the specific reasons why planning is important for AI projects:

- **Identify the problem:** Start by identifying a specific problem or opportunity that AI can help to solve or address. This could be anything from improving customer service to automating repetitive tasks to identifying new revenue streams. Lots of AI projects are force-fitting AI into a business flow even though there is no problem in the first place.

31

- **Define project goals and objectives:** Planning helps
  to ensure that the project has clear, measurable, and
  attainable goals and objectives. This helps to guide the
  development of the AI system and ensure that it aligns
  with the organization's overall strategy.

- **Identify and address potential risks:** Planning
  allows teams to identify and assess potential risks that
  may impact the project, such as lack of data, lack of
  expertise, or difficulty integrating the AI system into
  existing processes. Addressing these risks early on can
  help to mitigate them and ensure the project's success.

- **Build the right team:** Assemble a team with the right
  mix of skills, including data scientists, engineers,
  and domain experts. This will help to ensure that the
  project is executed effectively and that the AI system is
  tailored to the specific needs of the organization.

- **Manage expectations:** Planning helps to manage
  expectations by clearly communicating the project's
  goals, objectives, timeline, and budget to stakeholders.
  This helps to ensure that everyone is on the same page
  and that the project is delivered as expected.

- **Identify and address ethical considerations:**
  Planning allows teams to identify and address ethical
  considerations, such as the potential for bias in the data
  or the algorithms, that may impact the project.

- **Understand the data:** Make sure you have access to the
  necessary data and that it is of high quality. Understand
  how the data is collected, how it is stored, and how it
  can be used to train and operate the AI system.

- **Measure the results:** Plan and define the right KPIs and metrics to measure the results of the AI project. This will help to determine the system's effectiveness and avoid confusion later.

- **Communicate the value:** Communicate the value of the AI project to stakeholders, both inside and outside of the organization. This will help to build support and ensure that the project is seen as a success.

Overall, planning is important for AI projects because it helps to ensure that the project is well-defined, is well-organized, and has a clear path to success.

Approach

Approaching an AI project requires careful consideration of several key factors including the type of problem being addressed, the available data, the computing resources available, and the deployment requirements. Here are some general guidelines for approaching an AI project:

- **Model type:** Depending on the problem you are trying to solve, choose an appropriate machine learning or deep learning model. There are many different models to choose from, such as neural network–based models, decision trees, and support vector machines. Your choice of model will depend on the specific problem you are trying to solve, as well as the amount and quality of your data.

  - **Classification:** Use classification when you have labeled data and want to predict a categorical outcome. This is often used in scenarios where you want to identify which class or category a new data point belongs to. For example, you might use classification to predict whether a customer will buy a product or not based on their demographic information.

- **Regression:** Use regression when you have labeled data and want to predict a continuous outcome. This is often used in scenarios where you want to predict a numerical value, such as the price of a house or the amount of rainfall in a particular region. Regression models are useful when you want to make predictions that are not limited to a discrete set of values.

- **Unsupervised learning:** Use unsupervised learning when you have unlabeled data and want to identify patterns or groupings in the data. This is often used in scenarios where you want to discover hidden relationships or structures in your data, without having any prior knowledge about what you are looking for. Clustering is a common unsupervised learning technique that can be used to group similar data points based on their features.

It is also important to consider the size and quality of your data when deciding between these different types of learning. If you have a small dataset with limited features, you may want to use a simpler model like logistic regression or k-nearest neighbors. On the other hand, if you have a large dataset with complex features, you may want to use a more sophisticated model like a deep neural network.

Ultimately, the choice of learning algorithm will depend on the specifics of your project and your data.

- **Framework:** Select a framework for your AI project based on your choice of model. There are many popular frameworks available for building machine learning and deep learning models, including TensorFlow, PyTorch, and Scikit-learn. Choose a framework that is well-suited to your model and that has the necessary features and functionality you need.

- **Deployment details:** Deploying an AI model is an important part of the overall project and should be planned ahead of time to ensure that the model can be integrated smoothly into a production environment. Here are some steps you can take to plan the deployment of your AI model:

  - **Identify the deployment requirements:** Begin by identifying the requirements for deploying your model in a production environment. Consider factors such as the hardware and software infrastructure needed to run the model, the data inputs and outputs, and any security or compliance considerations.

  - **Select a deployment strategy:** Once you have identified the deployment requirements, choose a deployment strategy that is appropriate for your model and project goals. Some common strategies include deploying the model as a web service, integrating it into an existing application, or deploying it to an edge device.

- **Choose a deployment platform:** Depending on your deployment strategy, choose a deployment platform that is well-suited to your model and project goals. There are many platforms available for deploying AI models, including cloud platforms such as AWS, Azure, and Google Cloud, as well as edge devices such as the Raspberry Pi and Nvidia Jetson.

With a proper approach before even implementation, we can ensure that the AI projects will deliver value over the long term and don't end up as toy projects.

Approval Mechanism/ Organization Alignment

Establishing an approval mechanism for AI projects is an important step to ensure that the projects are aligned with organizational goals and values. Here are some steps to follow to establish an approval mechanism for AI projects:

1. **Identify stakeholders:** Start by identifying the stakeholders who should be involved in the approval process. This may include executive leadership, legal counsel, data privacy and security experts, and representatives from affected business units.

2. **Establish evaluation criteria:** Define the criteria that will be used to evaluate AI projects. This may include factors such as data quality, the potential impact on stakeholders, the value generated, and alignment with organizational goals.

3. **Develop an approval process:** Create a formal approval process that outlines the steps and stakeholders involved in the decision-making process. This may include submitting a project proposal, evaluating the proposal against the established criteria, and making a final decision.

4. **Implement a review board:** Establish a review board that is responsible for reviewing and approving or rejecting AI projects. The review board should include representation from the identified stakeholders and be responsible for ensuring that the project meets the established evaluation criteria.

5. **Conduct a risk assessment:** Before approving an AI project, conduct a thorough risk assessment to identify potential risks and develop mitigation strategies. This may involve assessing the potential impact on stakeholders, identifying potential biases in the data, and evaluating the accuracy and reliability of the AI model.

By establishing an approval mechanism for AI projects, you can ensure that AI projects are evaluated responsibly and ethically and that they align with organizational goals and values.

Key Performance Indicators

Measuring the results of an AI project is crucial for understanding its effectiveness and identifying areas for improvement. By using clear metrics, collecting and analyzing data, comparing data to baselines, identifying areas for improvement, and communicating the results, organizations can ensure that the AI system is aligned with the project's goals and objectives and continues to generate value over time.

Define Clear Metrics

Define clear and specific metrics that will be used to measure the results of the AI project. These metrics should align with the project's goals and objectives and be easy to understand and communicate.

Some examples of metrics that can be used to measure the results of an AI project include accurate churn rate for churn prediction model, customer satisfaction score, average delivery time for route optimization algorithm, ROI, etc.

Collecting the right key performance indicators (KPIs) is an important step in measuring the results of an AI project. KPIs are specific metrics that are used to measure the performance of a system or process, and they help to provide a clear and objective way to measure the results of an AI project.

When selecting KPIs for an AI project, it's important to ensure that they align with the project's goals and objectives and that they are easy to understand and communicate. It's also important to consider the data that will be required to measure the KPIs and make sure that the data is available and of high quality.

It's crucial to do the following:

- Track business KPI performance against benchmarks over time.

- Analyze business KPIs versus model KPIs for insights into how model performance is affecting the bottom line.

- Calculate realized value (in dollars) and the ROI.

Obtaining business KPIs from stakeholders is an important step in measuring the results of an AI project. Here are a few steps for getting these KPIs from stakeholders:

1. **Identify the stakeholders:** Identify the key
    stakeholders in the organization, including those
    who will be affected by the AI project, those
    who will be using the system, and those who
    will be responsible for its implementation and
    maintenance.

2. **Communicate the goals and objectives:** Clearly
    communicate the goals and objectives of the AI
    project to stakeholders. This will help to ensure that
    everyone is on the same page and that the project is
    aligned with the organization's overall strategy.

3. **Involve stakeholders in the planning process:**
    Involve stakeholders in the planning process, such
    as identifying the problem, understanding the data,
    and defining the use case. This will help to ensure
    that the project is tailored to the specific needs of
    the organization and that stakeholders are invested
    in its success.

4. **Identify the KPIs:** Work with stakeholders to
    identify the specific KPIs that will be used to
    measure the results of the AI project. This will help
    to ensure that the KPIs align with the project's goals
    and objectives and are relevant to the stakeholders.

5. **Communicate the process:** Communicate the
    process for collecting and analyzing data, as well
    as the schedule for measuring the results. This will
    help to ensure that stakeholders understand the
    process and can provide the necessary data and
    support.

6. **Measure the KPIs:** Measure the KPIs for AI projects as planned once models are in production and perform impact analysis based on controllable factors. Evaluate if AI is truly adding value and helping to make appropriate decisions.

7. **Keep stakeholders informed:** Keep stakeholders informed of the results of the AI project, including the performance of the system, areas for improvement, and any changes that have been made.

By following these strategies, organizations can obtain the necessary business KPIs from stakeholders and ensure that the AI project is aligned with the organization's goals and objectives.

Value

One of the biggest challenges that organizations face is to quantify the benefits of their AI initiatives. Value realization involves effort, and until one knows what benefits they are getting through their AI initiatives, they cannot improve them. No wonder most of the AI use cases fail to get any adoption at all due to the complexities in realizing the business value of the use cases.

As shown in Figure 2-2, processing data, building models with good accuracy, and deploying the models seems easier than tracking and realizing the value of AI. It seems difficult because we have come from a value standpoint when building AI projects and have never tracked the KPIs and return on investment.

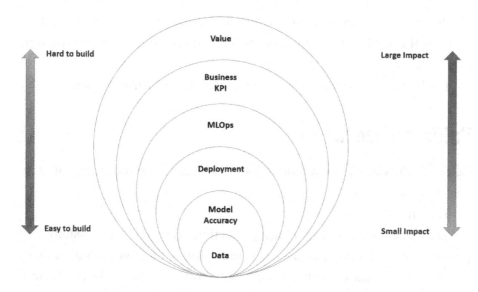

Figure 2-2. *Data to value*

Let's discuss what return on investment is and how to calculate it for an AI project.

Return on Investment

The *return on investment* (ROI) of the project should consider the costs of implementing the project and the potential value it will bring to the business.

The ROI for an AI project refers to the expected or realized financial return on investment for implementing an AI solution or initiative. It is determined by comparing the cost of implementing the project, including the cost of data acquisition, hardware and software, and personnel, to the expected or realized benefits.

The benefits of an AI project can be difficult to quantify but may include increased efficiency, improved accuracy, better decision-making, and reduced labor costs. The ROI for an AI project can be calculated by dividing the net financial benefits, such as increased revenue or reduced costs, by the total cost of the project and then multiplying the result by 100 to get a percentage.

But how do we calculate the potential value? Let's discuss that.

Value per Decision

Calculate the value per decision by estimating the potential impact of each decision made by the AI project.

Example 1:

Let's take the example of churn prediction again. If we successfully predict a customer is going to leave your business, the user can take some actions to retain the customer. So, the value from one customer would be X dollars.

In this scenario,

Value from one prediction = One generated from one customer

Example 2:

In some cases, we might want to know the overall value generated, not at prediction level.

The value per prediction can be calculated by dividing the total value of the project by the number of decisions it will make.

By considering these factors, you can derive the potential value from AI projects or calculate the value per decision.

Keep in mind that the actual value may vary based on various other factors, such as the following:

- **Probabilistic or uncertain outcomes:** Since the predictions are probabilistic, every prediction won't give us 100 percent value.

- **Cost of errors:** It's the cost of errors made by the model.

So, the overall ROI is as follows:

$$ROI = Return / Investment$$

where Return = Benefits from the model − Uncertainty of the predictions

Investments = Resources * Cost of resources

Keep in mind that deriving the dollar value from an AI prediction can be challenging, especially when the benefits are not directly financial. However, by identifying the problem, defining the use case, identifying the costs and benefits, calculating the ROI, and communicating the value, organizations can get a better understanding of the financial impact of the AI system.

Consumption of the AI Predictions

A major part of the AI life cycle is making sure the business consumes the AI predictions. Otherwise, the value won't be generated. So, while planning, it's crucial to understand how the business is going to consume the outputs. The following are the steps to be followed to have a clear understanding of this:

1. **Understand the prediction:** Before consuming an AI prediction, it's important to understand what it means and how it was generated. This includes understanding the input data, the algorithm used, the model's accuracy and limitations, and any assumptions made during the modeling process.

2. **Determine how to use the prediction:** Once you understand the prediction, determine how it can be used to make a decision or take action. This could involve using the prediction as input to another

system or using it to trigger an alert or notification. It should also include the following:

- **Explainability:** The system must be able to provide clear and transparent explanations for its decisions and recommendations, which helps to build trust and understanding among stakeholders.

- **Scalability:** The system must be able to handle large amounts of data and handle the complexity of the problem and adapt to changing conditions.

- **Flexibility:** The system should be able to handle different types of problems and be used in different areas of the organization.

- **Integration:** The system should be able to integrate with existing systems and processes to ensure a smooth and efficient workflow.

3. **Consider the confidence level:** AI models typically provide a level of confidence or probability associated with a prediction. It's important to consider this confidence level when deciding how to use the prediction. For example, if the confidence level is low, it may be necessary to verify the prediction with other data or methods before acting.

4. **Determine the appropriate storage solution:** Depending on the type of AI prediction and the use case, you may need to use different storage solutions. For example, if the prediction is an image or video, blob storage may be appropriate, while for text data, API calls may be more suitable.

5. **Set up the storage solution:** Once you have determined the appropriate storage solution, you will need to set it up and configure it appropriately. This may involve creating storage containers, configuring access policies and permissions, or integrating with other systems.

6. **Establish data flows:** To consume AI predictions, you will need to establish data flows between the AI model and the storage solution. This may involve setting up event triggers, establishing data transfer protocols, or configuring API endpoints.

7. **Ensure data security:** When working with sensitive data, it's important to ensure that appropriate security measures are in place. This may involve encrypting data in transit and at rest, configuring access controls and permissions, or integrating with other security systems.

8. **Identify user groups:** Once the problem has been defined, identify the different user groups who will be using the AI predictions. This may include different departments, teams, or individuals.

9. **Understand user needs:** For each user group, it's important to understand their specific needs and requirements for using the AI predictions. This includes understanding what decisions they need to make, what data they need to access, and what level of detail and accuracy they require.

10. **Provide training and support:** Once the AI predictions have been made available to users, it's important to provide training and support to ensure

that users can effectively consume and act on the
predictions. This may involve providing user guides,
training sessions, or support resources.

By following these steps, you can identify users for AI predictions and
ensure that they are provided with the information and resources they
need to effectively use the predictions to support decision-making and
improve outcomes.

Conclusion

In this chapter, we discussed the potential reasons for AI project failure,
the importance of decision intelligence in making better decisions using
AI, and the guidelines for the successful implementation of AI solutions.
It highlighted the need to ask critical questions before starting any AI
projects and debunked some of the myths surrounding AI. We also
introduced a DI requirements framework that covers planning, approach,
approval mechanism, KPIs, potential value, and consumption of AI
predictions, which are fundamental to AI projects.

CHAPTER 3

Decision Intelligence Methodologies

The idea of decision-making is the main topic of this chapter, along with its types. We will begin with the definition of decision-making and examine its various forms. Then, we will understand the decision-making process through an example. We will next go into greater detail about the development of decision-making, also known as *decision intelligence methodologies*, employed by humans over time, as well as their benefits and drawbacks. Our discussion will end with suggestions for decision intelligence approaches for various scenarios.

Decision-Making

To get to know the different decision intelligence methodologies, you have to understand what decision-making is.

Decision-making can be defined as the mental process of choosing among different options, culminating in the selection of a belief or action. Thus, decision-making consists of two major components.

- An objective function
- Alternatives/options available to reach the objective

A. Kulkarni et al., *Introduction to Prescriptive AI*,
https://doi.org/10.1007/978-1-4842-9568-7_3

The objective function is, in simple terms, the goal to achieve. Objectives can vary in complexity, from something as simple as selecting an outfit for a party to something as complex as finding a cure for cancer. The objective function typically has multiple ways to achieve it, known as *alternatives* or *options*, which must be evaluated to arrive at a decision. The decision-making process involves evaluating these alternatives to achieve the given objective. For example, when choosing an outfit for a party, a gentleman might have two options: a blue tuxedo and a black tuxedo. He might consider a variety of factors such as personal preference, what looks good, the theme of the party, etc., to make his decision, or he might randomly pick something without giving much thought to it. The thought process behind the decision can be complex or simple, and the decision-maker's reasoning, whether rational or irrational, is an important part of the process and impacts the outcome.

Decision-making can be classified in many ways, depending upon the factors taken into consideration such as single- and multiple-criterion decision-making, the number of people involved (individual and group decision-making), and the level of decision-making (strategic, tactical, and operational decision-making).

Types of Decision-Making

There are basically three types of decision-making. Let's examine each in detail.

Individual vs. Group Decision-Making

As the name suggests, individual decision-making is something that is the responsibility of a single person. This type of decision-making is mostly seen in the personal lives of individuals wherein they are responsible for making their own life decisions. It is a type of problem-solving that is used to identify and choose a course of action from among several alternatives.

In individual decision-making, the person is solely responsible for the outcome of the decision, and the decision is based on their own personal values, beliefs, and experiences.

Group decision-making, on the other hand, refers to the process of making a decision with a group of individuals. This type of decision-making can be used in various settings, such as in the workplace, in government, or in social organizations. Group decision-making is a way of problem-solving that is used to identify and choose a course of action from among several alternatives. The group members work together to identify and evaluate the different options and then make a decision that is collectively agreed upon. Group decision-making can be more effective than individual decision-making because it allows for the sharing of diverse perspectives and ideas and can increase the likelihood of better decisions being made.

Single- vs. Multiple-Criterion Decision-Making

Single-criterion decision-making (SCDM) is a decision-making approach where only one criterion is used to evaluate alternatives. In this type, the decision-maker selects the alternative that maximizes the value of the chosen criterion. This approach is useful when the decision is based on a single objective, such as maximizing profit/revenue or minimizing cost/time. Examples of SCDM include choosing an apartment for rent based on the price or buying shoes based on the brand.

Multiple-criterion decision-making (MCDM) is a decision-making approach where, unlike SCDM, multiple criteria are used to evaluate alternatives. This type of decision-making involves a trade-off between different criteria and often requires the decision-maker to assign weights to each criterion to reflect their relative importance. MCDM is useful when the decision involves multiple objectives, such as, for the previous example, choosing an apartment for rent not only based on the price but also based on location, size, view, distance to public transportation, etc.

One of the main advantages of MCDM over SCDM is that it provides a more comprehensive evaluation of alternatives. In MCDM, decision-makers can consider a range of factors and evaluate trade-offs between them. Additionally, MCDM can help to ensure that other important criteria are not overlooked, which can happen in SCDM when only one criterion is used to make the decision and the rest of them are ignored. However, MCDM can also be more complex and time-consuming than SCDM, and it requires more information and analysis. Additionally, assigning weights to criteria can be difficult, and the results may be sensitive to the weights chosen.

In summary, SCDM is appropriate when the decision is based on a single objective, while MCDM is more appropriate when the decision involves multiple objectives that need to be traded off against each other.

Strategic, Tactical, and Operational Decision-Making

Strategic, tactical, and operational decision-making is commonly observed in organizations. Figure 3-1 is a high-level overview of the decision-making process under strategic, tactical, and operational conditions. It explains what questions these types of decision-making answer, their execution time frame, and their frequency. Let's look at each one of them in detail.

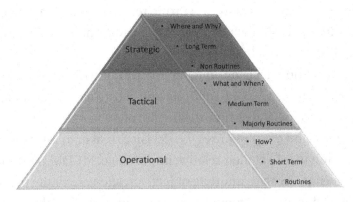

Figure 3-1. *Strategic, tactical, and operational decision-making*

Strategic decision-making is the process of determining the long-term goals and objectives of an individual/organization and the actions needed to achieve them. In an organizational setting, these types of decisions are usually made by top-level management and typically involve significant investments and changes to the organization's overall strategy and structure. Hence, they usually are nonroutine decisions. These types of decisions are usually slow moving, and changes happen over a long period of time.

Tactical decision-making involves the implementation and execution of medium-term plans and objectives. In an organizational setting, these types of decisions are usually formulated by middle-level management. These types of decisions often involve the management of day-to-day operations and the allocation of resources to achieve medium-term goals that align with the organization's overall strategy. These types of decisions are a mix of routine and nonroutine tasks and move faster than the strategic ones but are slower than that of the operational ones and are planned for medium to long term.

Operational decision-making is the process of making decisions related to the day-to-day management and operations of an individual/ organization. These decisions are usually made by front-line managers and employees and include scheduling, production, and other activities that are necessary to maintain the smooth functioning of the organization. These types of decisions are the fastest moving ones and are planned for the short term and involve routine tasks.

Decision-Making Process

As per the research by Baker et al. (2001), before making a decision, it is important to identify the individuals or groups that are responsible for making the decision and those who will be affected by it. This helps to reduce any possible disagreement about problem definition, requirements, goals, and criteria. Once the individual/group identification is done, it is recommended to follow a structured approach for decision-making, as shown in Figure 3-2.

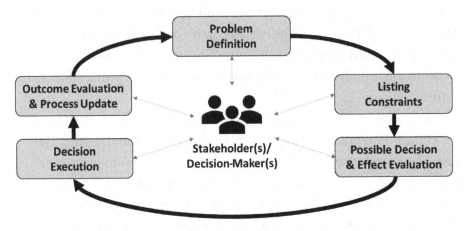

Figure 3-2. *Decision-making process*

Step 1: Problem Definition

This is the most critical and often ignored step in the decision-making process. Usually, framing the problem statement is a more iterative and time-consuming process than the solution itself. It requires input from the stakeholders, understanding the environment under which a decision has to be made, and what will happen if the problem is left as is, i.e., no decision is made. Once the problem definition is reviewed and agreed on by all will this step be considered done, and one can move to the subsequent steps.

The problem definition stage also requires the "desirable" state to be defined, i.e., what the ideal solution to the problem looks like once the decision is made.

A wrong problem definition will have wrong decisions made, and the actual problem will still remain the same, causing negative effects on the individuals/groups.

Step 2: Listing Constraints

Decision-making is, almost all the time, never done without constraints. These constraints can be financial, ethical/moral, environmental, time related, etc. The constraints help to filter out the alternatives and bring them from a possibly infinite number to a finite number. Listing all the constraints for a given problem is as important as the problem definition. Wrong constraints can lead to no or wrong decisions.

The constraints can be categorized into hard and soft ones. The hard constraints are the ones that are bound to be fulfilled and are non-negotiable in any condition. These constraints cannot be overlooked. The soft constraints, on the other hand, are the ones that are not "compulsory" but "good to have." They add more value to the decision-making but do not act as a blocker, if not fulfilled.

Step 3: Possible Decisions and Their Effects Evaluation

Once the problem is defined and constraints are listed, one needs to evaluate the possible decisions/solutions and their closeness to the desired state mentioned in step 1. Listing the pros and cons of every possible decision/solution within the scope of the constraints is a good practice to choose the right one.

Step 4: Decision Execution

Once all the possible solutions/alternatives are evaluated and the best one is chosen, it is time to execute the decision under the given constraints.

Step 5: Outcome Evaluation and Process Update

This step enables you to monitor and compare the outcome of the decision made to that of the expected one shown in step 3. This helps to re-evaluate the whole process and change the problem, constraints, and decisions if required.

These steps continue in a cyclic fashion until the right decision with the right outcome emerges. Let's go through the decision-making process with an example.

Decision-Making Process Example

In this example, we will go through the decision-making process of a person wanting to purchase a car. We will see how the person navigates the process, as listed in Figure 3-2, to arrive at the decision of which car to purchase based on their needs.

Step 1: Problem Definition

Amit, a 25-year-old unmarried software professional from Bangalore, is planning to purchase a car. It will be his first car, so he is super excited about it. He wants to ensure he does all the due diligence in the process to purchase a car so that he doesn't regret it later.

Step 2: Listing Constraints

Here are the hard constraints:

- **Car condition:** Brand new (not considering used cars)

- **Budget:** INR 8-10 lakhs ($9.6K–$12K)

- **Size:** Hatchback, 13.7 feet long (164.4 inches) × 6 feet wide (72 inches) × 5.4 feet tall (64.8 inches), comfortable for four people

- **Safety Ratings:** 4 or 5 (NCAP)

- **Purpose:** Travel to work and for short trips with family during weekends

- **Engine:** Diesel (manual), for better mileage

- **Boot space:** Minimum 250 liters

- **Waiting Period:** 1 to 3 months

- **After sales service charges:** Economical (INR 25–35K per year)

- **Authorized service centers:** Maximum coverage, even in remote areas

- **Comfort:** Comfortable and smooth driving quality

Here are the soft constraints:

- Premium interiors

- Cheap spare parts cost

- Wireless charger

- Top-notch music system

- Touchscreen display

- Alloy wheels

- Wide variety of color options

- Hill assist feature

- Good resale value

Step 3: Possible Decisions and Their Effects Evaluation

As this is the first time Amit is buying a car, he reaches out to his family and close friends, relatives, etc., who have a hatchback or have had one in the past. He wants to ensure at least the after service and authorized service center details are known through their experience. He also does secondary research with details available on social media, brand websites, forums, etc., that extend his scope of information beyond family/peers and validate their claims. He prepares a list of the constraints and the values/parameters according to the requirements (as shown in Figure 3-3). He also visits the showrooms of the brands to test-drive the vehicles and find out details such as drive quality, comfort, and waiting period.

Constraints	Type	Brand A	Brand B	Brand C	Brand D	Brand E
Price (in lakhs, Diesel and Topend Only)	Hard	8.98	9.83	8.05	7.41	6.33
Safety Rating	Hard	5	4	4	5	3
Engine Capacity (in cc)	Hard	1197	1199	1197	1197	1197
Mileage (in kmpl)	Hard	20	13	18	16	12
Boot Space Capacity (in litres)	Hard	268	318	345	341	279
Waiting Period (in months)	Hard	3	2	1	3	0.5
Average Maintainance Cost (in thousands)	Hard	30	25	24	32	33
No. of Authorized Service Centres	Hard	1644	1106	1035	768	679
Comfort	Hard	High	Low	High	Low	High
Drive Quality	Hard	High	Low	High	Medium	Low
Interior Quality	Soft	Premium	Basic	Premium	Basic	Basic
Spare Parts Cost	Soft	Cheap	Expensive	Cheap	Expensive	Expensive
Wireless charger	Soft	Available	Not Available	Available	Not Available	Not Available
Top notch music system	Soft	Available	Available	Available	Available	Available
Touchscreen display	Soft	Available	Available	Available	Available	Not Available
Alloy wheels	Soft	Available	Available	Available	Available	Available
Number of colour options	Soft	5	7	4	3	3
Hill Assist feature	Soft	Available	Available	Not Available	Not Available	Not Available
Current resale value (in lakhs)	Soft	4.19	5.93	3.98	5	3.2

Figure 3-3. *Possible decisions and their effects evaluation*

Step 4: Decision Execution

From Table 3-3, it seems like Brand A and Brand C are the closest in terms of what Amit needs. Brand A is a little costlier than Brand C, has less trunk space, has a longer waiting period, and has an average maintenance cost. Brand C's safety rating is lesser than that of Brand A, with slightly less mileage, fewer authorized service centers, fewer color options, and less resale value. After doing the cost-benefit analysis and immediate requirements, Amit goes ahead and buys Brand C. The price and waiting period played a bigger role in his decision-making than some of the other factors.

Step 5: Outcome Evaluation and Process Update

Ideally, for a once-in-a-lifetime type of decision, this step is not required as the decision-maker often makes the decision only after thorough research. However, Amit might face issues after buying the car, such as loan issues, hidden charges by the showroom, a sudden extension of the waiting period, etc. So, Amit might cancel his purchase and go for Brand A or re-evaluate all other options. Amit could also face issues post-delivery of the car, such as lower mileage than claimed, faulty car components, higher than expected maintenance cost, etc., which could force him to sell his car and consider buying a new one or look into used cars.

Decision-Making Methodologies

Humans have evolved over the years, and along with them, the way they make the decisions has also seen lots of changes. From making decisions as an individual randomly to group decision-makers who are rational to combining random and process-oriented and individual and group decision-makers, the decision-making process has been constantly increasing in its complexity.

The decision-making methodologies can be broadly divided into three major categories, human-only, human-machine, and machine-only, as shown in Figure 3-4.

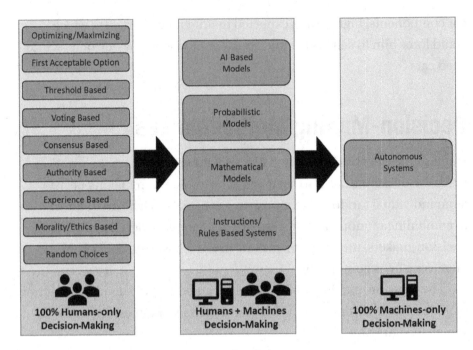

Figure 3-4. *Decision-making methodologies*

This type of categorization is purely based on the share of decision-making between humans and machines.

Let's take a closer look at decision-making methodologies in detail.

Human-Only Decision-Making

Human-only decision-making refers to the process of making choices or decisions solely on the judgment of human beings, without the involvement of artificial intelligence, machine learning, or any automated decision-making systems.

Before the advent of machines (from the abacus to computers and AI robots), humans have been making decisions either individually or in a group for personal, social, or organizational needs. The type of decisions that humans make vary depending upon the situation at hand, criticality of the decision, mental state of the people making the decision etc., ranging from totally random to outcome based to a more systematic complex decision-making.

The history of human decision-making can be traced back to the earliest civilizations, where decisions were made by leaders, elders, or individuals in positions of authority. Over time, the development of democratic institutions, laws, and norms has led to an increased emphasis on individual autonomy and the ability of individuals to make decisions for themselves.

Let's further evaluate the different types of human-only decision-making techniques.

Random Decisions

Random decision-making refers to the process of selecting an option or making a choice randomly, rather than relying on reasoning, intuition, or personal preference. This type of decision-making technique comes naturally to us and has been used in situations where the stakes are not high, there is an urgency to decide, there is no historical data to support the decision-making process, or all or any combination of these situations. Examples of random decision-making are flipping a coin or rolling a die.

Here are the advantages of random decisions:

- **Removes confirmation bias:** By relying on chance instead of personal preference, random decision-making can help to remove personal bias and increase objectivity.

- **Increases creativity:** Random selection can lead to unexpected and creative solutions.

- **Enhances exploration:** By forcing individuals to consider options they may not have otherwise considered, random decision-making can encourage exploration and increase the chances of discovering new and innovative solutions.

These are the disadvantages:

- **Lack of control:** Random decision-making leaves little room for control or direction, which can lead to suboptimal or even negative outcomes.

- **Inefficient:** It can be time-consuming to generate enough random outcomes to make an informed decision.

- **Unpredictable:** The outcomes of random decision-making can be unpredictable and difficult to control, leading to a lack of confidence in the decision-making process.

Morality/Ethics Based

Ethics-based or moral-based decision-making refers to the process of making choices and taking actions that align with a set of moral or ethical principles and values. The outcomes of these type of decisions may or may not be favorable to the decision-maker; however, they are of the highest ethics. The opposite of this is unethical decision-making. The decision-maker deliberately makes a decision that is not ethical and is most of the time favorable to them. There are several types of ethics-based decision-making, including the following:

- **Utilitarianism:** This is a theory of morality that advocates actions that foster happiness or pleasure and oppose actions that cause unhappiness or harm. An example of utilitarianism would be to choose playing with your kid over completing an office report, as you find more happiness in spending time with your kid.

- **Deontology:** This approach focuses on following a set of moral duties and obligations, regardless of the outcomes. A classic example of deontology would be "Don't lie. Don't steal. Don't cheat."

- **Virtue ethics:** This approach emphasizes the character and habits of the decision-maker, rather than rules or consequences, in determining the morality of a decision. An example of virtue ethics can be a customer going back to a store to settle the payment for an item that was unintentionally left unpaid or notifying the cashier if they issued a refund for an amount greater than what was owed.

Here are the advantages:

- **Consistency:** By following a set of ethical principles, individuals can make decisions that are consistent with their values and beliefs.

- **Increased trust:** Ethical decision-making can increase trust in organizations and relationships, as individuals perceive decisions as fair and just.

- **Personal fulfillment:** Making decisions that align with personal values and beliefs can lead to a sense of personal fulfillment and satisfaction.

Here are the disadvantages:

- **Subjectivity:** Ethics and morality can be subjective and open to interpretation, leading to disagreements and conflicts about the correct course of action.

- **Time-consuming:** Making decisions based on ethics can be time-consuming and require extensive research and consideration.

- **Potential for negative consequences:** Ethics-based decision-making may prioritize moral principles over practical considerations, leading to decisions with negative outcomes.

Experience Based

Experience-based decision-making refers to the process of making choices and making actions based on past experiences, knowledge, and expertise. There are several types of experience-based decision-making, including the following:

- **Rule-based:** This involves following established protocols, procedures, and guidelines based on past experiences.

- **Intuition-based:** This involves relying on gut feelings or instinct, based on past experiences and knowledge, to make decisions.

- **Evidence-based:** This involves making decisions based on data, research, and past experiences to inform and support the decision-making process.

Here are the advantages:

- **Speed:** Experience-based decision-making can be quicker, as individuals can rely on past experiences to inform their choices.

- **Improved accuracy:** Experience-based decision-making can lead to more accurate outcomes, as individuals can draw on past experiences to inform their choices.

- **Personalized:** Experience-based decision-making can be personalized, as individuals can tailor their decisions based on their own experiences and knowledge.

Here are the disadvantages:

- **Bias:** Past experiences can introduce bias into the decision-making process, leading to suboptimal or even negative outcomes.

- **Lack of innovation:** Relying solely on past experiences can limit the scope for innovation and limit the exploration of new solutions.

- **Inflexibility:** Experience-based decision-making can lead to inflexibility and a resistance to change, as individuals may be unwilling to deviate from established protocols or approaches.

Authority Based

Authority-based decision-making, aka "just follow orders" decision-making, is a process in which decisions are made by individuals in positions of authority, and subordinates are expected to carry out those decisions without question. This approach values obedience and hierarchy and is often used in situations where time is limited or immediate action is required. In such scenarios, the responsibility for decision-making is centralized, and individuals lower in the hierarchy are expected to carry out the decisions of their superiors. This approach can lead to efficient implementation of decisions, but it can also discourage initiative, creativity, and critical thinking and can create a lack of accountability for the outcomes of decisions. There are several types of authority-based decision-making, including the following:

- **Hierarchical:** This involves making decisions based on the chain of command within an organization or group.

- **Legal:** This involves making decisions based on laws, regulations, and legal mandates.

- **Expert-based:** This involves making decisions based on the advice or guidance of individuals with specialized knowledge or expertise.

Here are the advantages:

- **Efficiency:** Authority-based decision-making can be efficient, as individuals can rely on the guidance of higher authorities or experts to make decisions.

- **Clarity:** Authority-based decision-making can provide clarity, as individuals have a clear understanding of who is responsible for making decisions and what their role is.

- **Reduced responsibility:** Authority-based decision-making can reduce personal responsibility, as individuals can defer to the decisions of higher authorities or experts.

Here are the disadvantages:

- **Lack of autonomy:** Authority-based decision-making can limit personal autonomy and creativity, as individuals are limited by the decisions of higher authorities.

- **Potential for abuse:** Authority-based decision-making can lead to abuse of power, as individuals in positions of authority may abuse their power to make decisions that serve their own interests.

- **Inflexibility:** Authority-based decision-making can lead to inflexibility, as individuals may be limited in their ability to adapt to changing circumstances and make decisions that are best for the situation.

Consensus Based

Consensus-based decision-making is a group decision-making process in which all members of a group participate in reaching a decision that everyone can support. This approach values collaboration, communication, and agreement, rather than relying on a single person or a majority vote. Participants work together to identify and explore options and seek to find a solution that is satisfactory to all. The goal of consensus decision-making is to find a solution that addresses the concerns and needs of all parties involved, leading to more committed and effective implementation of the decision. There are several types of consensus-based decision-making, including the following:

- **Unanimity:** This involves making decisions based on the agreement of all individuals in a group.

- **Collaborative:** This involves making decisions through a collaborative process, in which individuals work together to reach a shared understanding and agreement.

These are the advantages:

- **Inclusiveness:** Consensus-based decision-making can promote inclusiveness, as all individuals in a group have a voice and a role in the decision-making process.

- **Improved outcomes:** Consensus-based decision-making can lead to improved outcomes, as individuals can draw on the diverse perspectives and experiences of the group to inform their choices.

- **Increased commitment:** Consensus-based decision-making can increase commitment to decisions, as individuals are more likely to support and implement decisions that they have helped to shape.

These are the disadvantages:

- **Time-consuming:** Consensus-based decision-making can be time-consuming, as individuals may need to engage in extensive discussions and negotiations to reach a shared agreement.

- **Potential for groupthink:** Consensus-based decision-making can lead to groupthink, in which individuals conform to the opinions of the group and suppress dissenting views.

- **Difficulty reaching agreement:** Consensus-based decision-making can be difficult, as individuals may have differing opinions, values, and experiences that make it challenging to reach a shared agreement.

Voting Based

Voting-based decision-making is a process in which a group of people cast ballots to determine a course of action. The decision is based on a tally of the votes, with a majority or supermajority needed to determine the outcome. This method is often used when time is limited or when a consensus cannot be reached. Voting can take various forms, such as a simple majority, a two-thirds majority, or a weighted vote based on individual member's stake or influence.

This approach values efficiency and fairness and ensures that the decision reflects the will of the group. However, it can also lead to minority views being ignored or overlooked. There are several types of voting-based decision-making, including the following:

- **Simple majority:** This involves making decisions based on the will of the majority of individuals in a group.

- **Supermajority:** This involves making decisions based on the agreement of a specified percentage of individuals in a group.

- **Plurality:** This involves making decisions based on the candidate or option with the most votes, even if they do not have a majority.

These are the advantages:

- **Fairness:** Voting-based decision-making can promote fairness, as all individuals in a group have an equal opportunity to influence the outcome.

- **Democracy:** Voting-based decision-making is a cornerstone of democratic systems, as it allows for the will of the people to be expressed and implemented.

- **Transparency:** Voting-based decision-making can promote transparency, as the results of a vote are clear and easily understood.

These are the disadvantages:

- **Limited representation:** Voting-based decision-making may not accurately represent the views of all individuals in a group, as some may not participate in the vote or may have their views diluted by the majority.

- **Biased outcomes:** Voting-based decision-making can lead to biased outcomes, as the results may be influenced by factors such as voter turnout or unequal representation.

- **Lack of accountability:** Voting-based decision-making can lead to a lack of accountability, as individuals may be less likely to take responsibility for the outcome of a vote or election.

Threshold Based

This type of decision-making takes certain thresholds into consideration for given activities to perform an action. This happens when the outcome sample space is very limited and there is a lot of historical information on the decisions taken based on the outcome. There are several types of threshold-based decision-making, including the following:

- **Quorum:** This involves making decisions based on the minimum number of individuals who must be present or represented in a group to make decisions.

- **Approval threshold:** This involves making decisions based on a minimum level of approval or support that must be met before a decision can be taken.

- **Performance threshold:** This involves making decisions based on a minimum level of performance or achievement that must be met before a decision can be taken.

These are the advantages:

- **Objectivity:** Threshold-based decision-making can provide objective standards for making decisions, as it is based on specific, measurable criteria.

- **Transparency:** Threshold-based decision-making can promote transparency, as the criteria for making decisions are clear and easily understood.

- **Consistency:** Threshold-based decision-making can lead to consistency in decision-making, as individuals are held to the same standards and expectations.

These are the disadvantages:

- **Rigidity:** Threshold-based decision-making can be rigid and inflexible, as it may not allow for consideration of contextual factors or individual circumstances.

- **Lack of creativity:** Threshold-based decision-making can limit creativity and innovation, as individuals may focus solely on meeting the minimum criteria rather than exploring new and innovative solutions.

- **Inadequate representation:** Threshold-based decision-making may not accurately represent the views and needs of all individuals in a group, as some may not meet the minimum criteria or may have their views diluted by others who do.

First Acceptable Match Based

First acceptable match, aka *satisficing*, decision-making is a process in which individuals choose the first option that meets a minimum threshold of acceptability, rather than searching for the best possible solution. This approach values efficiency and speed and is often used when decision-makers face time constraints or limited resources. In a satisficing approach, individuals set a minimum acceptable level of quality or performance and then choose the first option that meets that criterion. This method can lead to quick decision-making, but it may result in suboptimal outcomes, as it may overlook better options that require additional effort or resources to identify. Satisficing can also result in a failure to fully explore all available options and to consider the long-term consequences of decisions. There are several types of satisficing-based decision-making, including the following:

- **Bounded rationality:** This involves making decisions based on limited information, resources, and time, and choosing the first option that meets a minimum level of acceptability.

- **Satisficing heuristics:** This involves making decisions based on rules of thumb or shortcuts that help individuals quickly identify options that meet a minimum level of acceptability.

These are the advantages:

- **Efficiency:** Satisficing-based decision-making can be an efficient way to make decisions, as it allows individuals to quickly identify options that meet a minimum level of acceptability.

- **Realism:** Satisficing-based decision-making can promote realism, as it acknowledges that individuals have limited information, resources, and time, and that it is not always possible to achieve the best possible solution.

- **Practicality:** Satisficing-based decision-making can be a practical way to make decisions, as it focuses on what is achievable and realistic, rather than what is ideal or optimal

These are the disadvantages:

- **Lack of quality:** Satisficing-based decision-making may result in lower-quality outcomes, as individuals may settle for an option that meets a minimum level of acceptability, rather than striving for the best possible solution.

- **Limited perspective:** Satisficing-based decision-making may be based on limited information, resources, and time, and may not consider all relevant factors or perspectives.

- **Missed opportunities:** Satisficing-based decision-making may result in missed opportunities, as individuals may settle for an option that meets a minimum level of acceptability, rather than exploring alternative or better options.

Optimization/Maximization Based

Optimizing or maximizing decision-making is a process in which individuals strive to make the best possible decision based on all available information and constraints. This approach values finding the best solution and is often used when the consequences of a decision are significant or long-lasting. In an optimizing approach, individuals consider all available options, weigh the pros and cons, and choose the option that provides the greatest benefits or the least drawbacks. This method can lead to the best possible outcome, but it can also be time-consuming and resource-intensive, as it requires a thorough analysis of all options. Maximizing decision-making can also result in decision paralysis if individuals become overwhelmed by the available information and are unable to make a decision. There are several types of optimization/maximization-based decision-making, including these:

- **Linear programming:** This involves using mathematical models to optimize decisions based on constraints and objectives.

- **Nonlinear programming:** This involves using mathematical models to optimize decisions based on complex, nonlinear relationships between variables and constraints.

These are the advantages:

- **Quality:** Optimization/maximization-based decision-making can result in higher-quality outcomes, as individuals strive to find the best possible solution.

- **Efficiency:** Optimization/maximization-based decision-making can be an efficient way to make decisions, as it involves evaluating all options and choosing the best one.

- **Complete perspective:** Optimization/maximization-based decision-making can consider all relevant factors and perspectives, providing a complete and comprehensive perspective on the decision-making process.

These are the disadvantages:

- **Complexity:** Optimization/maximization-based decision-making can be complex, as it involves evaluating all options and considering multiple factors.

- **Resource-intensive:** Optimization/maximization-based decision-making can be resource-intensive, requiring significant time, information, and computational resources to find the best possible solution.

- **Unrealistic expectations:** Optimization/maximization-based decision-making can result in unrealistic expectations, as individuals may strive for the best possible solution, even if it is not achievable or practical.

With human-only decision-making, a lot of different cognitive biases tend to exist. Let's deep dive into the concept of cognitive bias and its different types.

Cognitive Bias Due to Human-Only Decision-Making

Cognitive bias refers to systematic errors in our thinking that can lead to flawed decisions. These biases arise from our emotions, experiences, and beliefs, and they can distort our perception of reality. Some common cognitive biases include confirmation bias, where we tend to favor information that supports our existing beliefs, and negativity bias, where we give more weight to negative information. Overconfidence bias, where we overestimate our abilities and the accuracy of our judgments, is another example. Understanding these biases is important, as they can impact our decision-making and lead to suboptimal outcomes. By recognizing and correcting for our biases, we can make more informed and objective decisions.

We will study cognitive biases in detail in the following chapters.

Human-Machine Decision-Making

As humans evolved, their decision-making techniques evolved too. Just like we invented tools/techniques to help ease our day-to-day life (e.g., wheels for traveling, tools for farming, etc.), we also started to create tools and techniques to aid our decision-making process. This led to human-machine interactions that helped us make decisions quickly and with more accuracy and to some extent mitigate the cognitive bias effects. Let's evaluate the different types of human-machine decision-making, aka machine assisted decision-making.

Instruction/Rule-Based Systems

In this technique, there is a computerized system that uses a set of explicitly stated rules to support decision-making activities. These rules are derived from domain expertise, regulations, or best practices, and they are encoded in the system to assist with decision-making processes. The system operates by matching the facts of a particular case against the rules and suggesting the course of action that best fits the rules. Rule-based human-machine decision-making is useful in situations where the decision-making process is well defined and the rules are well understood, as it provides a consistent and repeatable approach to decision-making. However, it may not be well suited to situations where the rules are complex or subject to change, as this requires manual updates to the system's rule base. Before the advent of computer systems, there were other tools used for decision-making. For example, the abacus was used for calculating finances. There are several types of instruction/rule-based decision systems, including the following:

- **Expert systems:** These are computer programs that use a set of rules and knowledge to mimic the decision-making abilities of human experts.

- **Decision trees:** These are graphical representations of a set of rules or conditions that specify how decisions should be made based on a set of variables or factors.

These are the advantages:

- **Consistency:** Instruction/rule-based decision systems can provide consistent outcomes, as they rely on a set of predefined rules that specify how decisions should be made.

- **Speed:** Instruction/rule-based decision systems can make decisions quickly, as they rely on predefined rules and algorithms.

- **Accuracy:** Instruction/rule-based decision systems can provide accurate outcomes, as they rely on well-defined rules and algorithms that have been tested and validated.

These are the disadvantages:

- **Rigidity:** Instruction/rule-based decision systems can be rigid, as they rely on a set of predefined rules that may not adapt to changing circumstances or new information.

- **Lack of flexibility:** Instruction/rule-based decision systems may not be flexible, as they rely on predefined rules that may not account for all possible scenarios or exceptions.

- **Limited perspective:** Instruction/rule-based decision systems may have a limited perspective, as they rely on predefined rules that may not consider all relevant factors or perspectives.

Mathematical Models

In this technique, there is a computer-based system that uses mathematical and statistical techniques to support decision-making. It integrates mathematical models, data, and algorithms to analyze complex problems and provide recommendations or solutions. The models used in a mathematical decision system can be linear or nonlinear programming, dynamic programming, or any other mathematical method suitable for a particular problem. The system can also include data visualization tools to help users understand the results of the analysis. The goal of a mathematical model is to provide objective and quantifiable insights into decision problems and to help users make more informed decisions. Mathematical modeling is commonly used in fields such as finance, engineering, and healthcare to support complex decision-making processes. There are several types of mathematical model-based decision systems, including the following:

- **Linear programming:** This involves using mathematical models to optimize decisions based on constraints and objectives.

- **Nonlinear programming:** This involves using mathematical models to optimize decisions based on complex, nonlinear relationships between variables and constraints.

- **Decision analysis:** This involves using mathematical models to evaluate different courses of action based on their expected outcomes, risks, and uncertainties.

These are the advantages:

- **Accuracy:** Mathematical models can provide accurate predictions about the outcomes of different decisions.

- **Speed:** Mathematical models can make predictions quickly, as they rely on algorithms and mathematical equations.

- **Consistency:** Mathematical models can provide consistent outcomes, as they rely on well-defined algorithms and equations that have been tested and validated.

These are the disadvantages:

- **Complexity:** Mathematical models can be complex, as they involve mathematical equations and algorithms that may not be easy to understand or interpret.

- **Limited perspective:** Mathematical models may have a limited perspective, as they rely on predefined relationships between variables that may not account for all relevant factors or perspectives.

- **Model assumptions:** Mathematical models rely on certain assumptions about the relationships between variables, which may not always hold true in reality. This can lead to inaccuracies in the predictions made by mathematical models.

Probabilistic Models

In this technique, there is a computerized system that uses probability theory to support decision-making. It is based on the idea that many real-world problems involve uncertainty and that a probabilistic approach is more appropriate than a deterministic one in such cases. The system uses probabilistic models, such as Bayesian networks or Markov decision processes, to represent the relationships between variables and the uncertainty associated with them. The models are used to calculate the probability of different outcomes given certain inputs and to support decision-making by providing recommendations or suggesting actions with the highest expected value. Probabilistic decision-making is commonly used in fields such as finance, insurance, and healthcare, where decisions must be made in the face of uncertainty. By explicitly modeling and quantifying uncertainty, a probabilistic model can provide more robust and flexible support for decision-making. There are several types of probability-based decision systems, including the following:

- **Bayesian networks:** These are probabilistic graphical models that represent the relationships between variables and their probabilities.

- **Monte Carlo simulation:** This involves generating many random samples from a probabilistic model to estimate the probabilities of different outcomes.

- **Decision analysis:** This involves using probability models to evaluate different courses of action based on their expected outcomes and risks.

These are the advantages:

- **Consideration of uncertainty:** Probability-based decision systems account for uncertainty by using probabilistic models to estimate the likelihood of different outcomes.

- **Accuracy:** Probability-based decision systems can provide accurate predictions about the likelihood of different outcomes, as they rely on statistical methods and algorithms.

- **Consistency:** Probability-based decision systems can provide consistent outcomes, as they rely on well-defined algorithms and equations that have been tested and validated.

These are the disadvantages:

- **Complexity:** Probability-based decision systems can be complex, as they involve statistical methods and algorithms that may not be easy to understand or interpret.

- **Limited data:** Probability-based decision systems rely on historical data or other relevant information to estimate the probabilities of different outcomes. If this data is limited or not representative, the predictions made by probability-based decision systems may not be accurate.

- **Model assumptions:** Probability-based decision systems rely on certain assumptions about the relationships between variables, which may not always hold true in reality. This can lead to inaccuracies in the probabilities estimated by probability-based decision systems.

AI-Based Models

In this technique, there is a computer system that uses artificial intelligence and machine learning algorithms to support decision-making. It analyzes data, predicts outcomes, and makes recommendations to help decision-makers make informed choices. It can be used in various applications such as assessing risk, optimizing business processes, and predicting future trends. The goal is to provide accurate and relevant information to decision-makers in real time, enhancing the speed and quality of decision-making. There are several types of AI or machine learning–based decision systems, including the following:

- **Supervised learning:** This involves training machine learning algorithms on labeled data to predict outcomes based on that data.

- **Unsupervised learning:** This involves training machine learning algorithms on unlabeled data to identify patterns and relationships within the segments of the data.

These are the advantages:

- **Automation:** AI or machine learning–based decision systems can automate decision-making processes, reducing the time and effort required for manual decision-making.

- **Accuracy:** AI or machine learning–based decision systems can provide accurate predictions and decisions based on the data they have been trained on.

- **Scalability:** AI or machine learning–based decision systems can be easily scaled up to handle large amounts of data, making them well-suited for large, complex decision-making processes.

These are the disadvantages:

- **Bias:** AI or machine learning–based decision systems can be biased if the data they have been trained on is biased or unrepresentative.

- **Explainability:** AI or machine learning–based decision systems can be difficult to interpret or explain, as their decisions are based on complex algorithms and mathematical models.

- **Limitations of training data:** AI or machine learning–based decision systems rely on the data they have been trained on to make predictions and decisions. If the training data is limited or not representative, the decisions made by AI or machine learning–based decision systems may not be accurate.

Machine-Only Decision-Making

The human-machine decision-making, over the years, led to humans realizing that some of the decision-making can be fully automated through machines, with little to almost no oversight needed. This led to a new, more advanced type of decision-making known as machine-only decision-making, and the machines that make those type of decisions are called *autonomous systems*. Let's go through a few examples of machine-only decision-making.

Autonomous Systems

Autonomous systems can make decisions and take actions without human intervention. They use artificial intelligence, machine learning, and other advanced technologies to analyze data, identify patterns, and make predictions. Autonomous systems are designed to operate independently and make decisions based on their understanding of the data and the situation. These systems can be used in various applications, such as robotics, unmanned aerial vehicles (UAVs), autonomous vehicles, and smart homes. The goal of autonomous systems is to provide real-time, accurate, and relevant information to support decision-making, while reducing the need for human intervention and increasing efficiency and speed. There are several types of fully autonomous decision systems, including the following:

- **Rule-based systems:** These systems make decisions based on preprogrammed rules, such as if-then statements.

- **Model-based systems:** These systems use mathematical, probabilistic, or AI-based models to make decisions, such as decision trees or neural networks.

- **Reinforcement learning systems:** These systems make decisions based on trial and error, with the goal of maximizing a reward signal.

These are the advantages:

- **Efficiency:** Fully autonomous decision systems can make decisions faster and more accurately than humans, as they are not subject to human biases or limitations.

- **24/7 operation:** Fully autonomous decision systems can operate 24/7 without the need for rest, making them ideal for tasks that require continuous operation.

- **Increased safety:** In some cases, fully autonomous decision systems can make safer decisions than humans, as they can process more information and make decisions based on that information in real time.

These are the disadvantages:

- **Lack of flexibility:** Fully autonomous decision systems can be limited by the rules, algorithms, or models they have been programmed with and may not be able to adapt to new situations or changing conditions.

- **Responsibility:** It can be difficult to determine who is responsible for the decisions made by fully autonomous decision systems, as they are not subject to human control or oversight.

- **Lack of transparency:** Fully autonomous decision systems can be difficult to understand, interpret, or explain, as their decisions are based on complex algorithms and mathematical models. This can make it difficult for humans to understand the reasoning behind the decisions made by fully autonomous decision systems.

Conclusion

As we have seen, the decision-making process and its different methodologies are complex. The structure of human society, comprising individuals, groups, organizations, governments, etc., add to the complexity. Hence, there usually is no single way of making decisions. It can involve a combination of different types of decision-making techniques with different methodologies. For example, a person planning to purchase an apartment can do so just by doing the following:

- Buying an apartment in the same society that they might be staying in, because they had a good past experience: strategic humans-only individual decision-making with single criterion, i.e., society

- Buying an apartment after researching online, asking friends/family/peers, and exploring different localities based on price, area, connectivity to public transport and workplace: strategic human-machine group decision-making with multiple criteria

This complexity of decision-making is what makes us human and has helped us to be on the top of the food chain. We can expect the decision-making process to become more complex over time, ensuring humans evolve further.

CHAPTER 4

Interpreting Results from Different Methodologies

In the previous chapter, we saw different decision intelligence methodologies, with their advantages and limitations. In this chapter, we will implement various human-machine decision-making methodologies using the Python programming language. Because covering all the methodologies is not possible in a single chapter, or even book, we will focus on the most commonly used ones: the mathematical, probabilistic, and AI-based techniques. Also, we will look at how to interpret the results generated from those methodologies and how to use them to make decisions.

Decision Intelligence Methodology: Mathematical Models

Mathematical models have long been used in various processes. As discussed in the previous chapter, these types of models fare better than the others because of their accuracy, execution speed, and consistency in the outcomes produced. These types of models perform very well in situations where the factors/variables have a consistent behavior and there

© Akshay Kulkarni, Adarsha Shivananda, Avinash Manure 2023
A. Kulkarni et al., *Introduction to Prescriptive AI*,
https://doi.org/10.1007/978-1-4842-9568-7_4

is less uncertainty involved. Hence, we can find a lot of use cases in the physics and engineering domains for mathematical models. Let's look at some examples of the linear and nonlinear mathematical models.

Linear Models

Linear models, as the name implies, are mathematical equations that describe a linear relationship between two or more variables. They are represented by the following equation:

$$y = \beta_0 + \beta_1 x_1 + \beta_2 x_2 + \ldots + \beta_n x_n$$

where:

y is the dependent variable (a variable that needs to be calculated/predicted based on the input).

$x_1, x_2, \ldots, x_n$ are the n independent variables used as input.

β_0 is a constant value. It is equal to y when $x_1, x_2, \ldots, x_n$ are all 0.

$\beta_1, \beta_2, \ldots, \beta_n$ are beta coefficients of the independent variables $x_1, x_2, \ldots, x_n$, respectively.

The linear model equations usually operate under certain constraints for the independent variables. Let's explore linear models through a real-world use case.

Problem Statement:

A factory produces four different products: diapers, sanitary pads, facial tissue, and toilet paper. The daily produced number of these four products is $x_1, x_2, x_3,$ and x_4, respectively.

Conditions:

Based on the following conditions, the factory management wants to come up with a production schedule to maximize profit:

- The profit per unit of product is $10, $11, $8, and $12.5 for diapers, sanitary pads, facial tissue, and toilet paper, respectively.

- Because of workforce constraints, the total number of units produced per day can't exceed 45.

- The raw materials required for producing the aforementioned products are wood pulp and fiber. To produce a single unit of a diaper, 5 units of wood pulp are consumed. Each unit of sanitary pads requires 7 units of wood pulp and 2 units of fiber. Each unit of facial tissue needs 2 units of wood pulp and 1 unit of fiber. Finally, each unit of toilet paper requires 3 units of fiber.

- Because of the transportation and storage constraints, the factory can consume up to 100 units of wood pulp and 75 units of fiber per day.

Solution:

As the problem is related to production optimization, the environment can be safely assumed to be a controlled one with little to no uncertainty. Hence, we can use a mathematical model as a solution.

Based on the given problem statement and associated conditions, the mathematical model can be defined like this:

Maximize: $10x_1 + 11x_2 + 8x_3 + 12.5x_4$ (Profit)

Constraints:

$x_1 + x_2 + x_3 + x_4 \leq 45$ (Manpower)

$5x_1 + 7x_2 + 2x_3 \leq 100$ (Wood pulp)

$2x_2 + 1x_3 + 3x_4 \leq 75$ (Fiber)

$x_1, x_2, x_3, x_4 \geq 0$

We will be using a general-purpose and open-source linear programming modeling package in Python named PuLP. It is a highly suitable option for a diverse set of optimization problems due to its exceptional qualities such as user friendliness, versatility, compatibility with other Python libraries, continual support from a thriving community, and open-source availability.

Step 1: Importing the PuLP library and defining the mathematical model object

```
[In]: from pulp import LpProblem, LpMaximize, LpVariable,
LpStatus, lpSum
[In]: opt_model = LpProblem(name="production-planning",
sense=LpMaximize)
```

Here the model object is defined using the LpProblem module with sense set to LpMaximize as we want to maximize the profit made from the production.

Step 2: Defining the decision variables

```
[In]: x = {i: LpVariable(name=f"x{i}", lowBound=0) for i in
range(1, 5)}
[In]: print(x)
[Out]: {1: x1, 2: x2, 3: x3, 4: x4}
```

Step 3: Adding constraints

```
[In]: opt_model += (lpSum(x.values()) <= 45, "Manpower")
[In]: opt_model += (6 * x[1] + 4 * x[2] + 2 * x[3] <= 100,
"Wood Pulp")
[In]: opt_model += (2 * x[2] + 4 * x[3] + 6 * x[4] <= 75,
"Fiber")
```

Step 4: Setting the objective function

```
[In]: opt_model += (10 * x[1] + 11 * x[2] + 8 * x[3] + 12.5 *
x[4], "Objective Function")
```

Step 5: Utilizing the default solver method for solving the given optimization problem

```
[In]: status = opt_model.solve()
```

Step 6: Getting the results

```
[In]: print(f"status: {opt_model.status}, {LpStatus[opt_model.
status]}")
[In]: print(f"objective: {opt_model.objective.value()}")
[In]: for var in x.values():
            print(f"{var.name}: {var.value()}")
[In]: for name, constraint in opt_model.constraints.items():
            print(f"{name}: {constraint.value()}")

[Out]:
status: 1, Optimal
objective: 512.5
x1: 20.0
x2: 0.0
x3: 0.0
x4: 25.0
Manpower: 0.0
Wood_Pulp: 0.0
Fiber: 0.0
```

From the previous output, we can see that the model suggests producing 20 units of x_1, i.e., diapers, and 25 units of x_4, i.e., toilet paper, for a maximum profit (\$512.5) under the given constraints. Now, in a dynamic real-world factory setting, the constraints will change, and hence there arises a need to build a solution that can update the constraints based on the changes (e.g., changes in profit value of the products) happening over time due to multiple factors. So, this solution can be built into an

API solution that can accept inputs (constraints and other details) either manually through an UI or through other system outputs, and the results can be consumed either through the same UI or through any other application (Excel, custom applications, etc.).

Figure 4-1 shows how the high-level solution architecture would look.

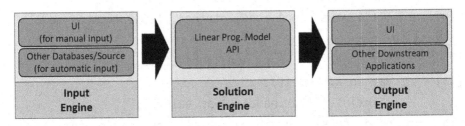

Figure 4-1. *High-level architecture: linear mathematical models*

Nonlinear Models

Linear models assume that the relationship between the inputs and outputs is always linear. This assumption is often not true in real-world situations where the relationship between variables is complex and nonlinear. This is where the need for nonlinear models arises. They can capture the complex behavior/relationships between the inputs and outputs more accurately. They can be represented by the following equation:

$$y = f(x)$$

where:

y is the dependent variable (a variable that needs to be calculated/predicted based on input).

f(x) could be any nonlinear function, such as a polynomial, exponential function, trigonometric function, or any other nonlinear function.

Let's understand how to solve a nonlinear problem with the help of the following use case.

Problem Statement:

After the COVID-19 pandemic, company X plans to gradually have its employees return to work at the office while maintaining a minimum of 25 percent occupancy in both of its office blocks to ensure efficient operations. The government has provided a risk index to assist companies in determining the risks associated with re-opening offices during the current situation. The risk index (RI) is defined in the following manner:

$$RI = 3x_1^2 + 3x_1^3 + x_1x_2 - 2x_2^2 + 2x_2^3$$

where:

x_1 = Proportion of block A available capacity

x_2 = Proportion of block B available capacity

What is the safest way (according to the RI) for company X to resume work from the office?

Solution:

With the RI definition and considering company X wants to minimize the risk, we can define the objective function as follows:

Minimize: $RI = 3x_1^2 + 3x_1^3 + x_1x_2 - 2x_2^2 + 2x_2^3$

Constraints:

$x_1 \geq 0.25$

$x_2 \geq 0.25$

$x_1, x_2 \leq 1$

For this nonlinear problem, we will use the scipy package to derive the solution.

Step 1: Importing the required libraries (especially scipy)

[In]:
```
import numpy as np
```

```
import pandas as pd
import random
from scipy.optimize import minimize
```

Step 2: Create a function to generate starting points

We will use the following function to generate a tuple of starting points within the feasible range for this problem:

```
[In]:
def geneate_kickoff_points(number_of_points):
    '''
    number_of_points [list]: how many points to generate
    '''
    kickoff_points = []
    for point in range(number_of_points):
        kickoff_points.append((random.random(), random.
        random()))
    return kickoff_points
```

Step 3: Formulating the objective function and constraints

```
objective_function = lambda x: (3*x[0]**2) + (3*x[0]**3) +
(x[0]*x[1]) - (2*x[1]**2) + (2*x[1]**3)
constraints = [
    {'type': 'ineq', 'fun': lambda x: x[0] - 0.25}, # Block A
    capacity >= 0.25
    {'type': 'ineq', 'fun': lambda x: x[1] - 0.25} # Block B
    capacity >= 0.25
]
boundaries = [(0,1), (0,1)]
```

Step 4: Running the optimal solution function

```
# generate a list of N potential kickoff points
[In]: kickoff_points = geneate_kickoff_points(20)
```

```
[In]: first_iteration = True

[In]:
for point in kickoff_points:
    # for each point run the algorithm
    result = minimize(
        objective_function,
        [point[0], point[1]],
        method='SLSQP',
        bounds=boundaries,
        constraints=constraints
    )
    # first iteration always going to be the best so far
    if first_iteration:
        better_solution_found = False
        best_solution = result
    else:
        # if we find a better solution, lets use it
        if result.success and result.fun < best_solution.fun:
            better_solution_found = True
            best_solution = result

# print results if algorithim was successful
[In]:
if best_solution.success:
    print(f"""Optimal solution found:
    - Proportion of block A available capacity: {round
      (best_solution.x[0], 3)}
    - Proportion of block B available capacity: {round
      (best_solution.x[1], 3)}
    - Risk index value: {round(best_solution.fun, 3)}""")
else:
```

```
    print("No solution found to problem")
[Out]:
Optimal solution found:
        -   Proportion of block A available capacity: 0.25
        -   Proportion of block B available capacity: 0.597
        -   Risk index value: 0.096
```

From the previous simulation, we find that company X can resume work from office for its employees with block A capacity at 25 percent and block B capacity at about 60 percent, with the optimal about 10 percent risk for the given constraints/conditions.

This solution can be deployed the same way as how the linear model was deployed. However, certain details such as the infrastructure might need changes as usually nonlinear solutions require more compute power for calculations, as compared to the linear ones.

Decision Intelligence Methodology: Probabilistic Models

Probabilistic models employ probability theory principles, unlike mathematical models that use mathematical equations, to model problems. This makes the probabilistic models particularly appropriate for situations where numerous uncertain factors influence the outcome. Probabilistic models are often used in fields such as finance, economics, and healthcare, where the outcomes of decisions are uncertain and unpredictable. We will now see how we can use one of the widely used probabilistic models, a Markov chain, to build a decision intelligence system.

Markov Chain

Markov chain models are a type of probabilistic models that can be employed to study dynamic systems, where the likelihood of transitioning from one state to another is determined solely by the present state, without any dependence on the system's past history.

An example of a Markov chain model is a stock market model that has three states: bull, bear, and flat. We can represent this as a three-state Markov chain model, where the transition probabilities between states depend only on the current state. For example, let's say that the probability of moving from a bull market to a bear market is 0.3, and it is 0.6 for staying in bull market itself. The probability of moving from a bear market to a flat market is 0.2, and it is 0.3 for staying in bear market itself. The probability of moving from a flat market to a bull market is 0.2, and the probability of staying in the flat market is 0.6. Figure 4-2 shows what the transition matrix for the stock market model.

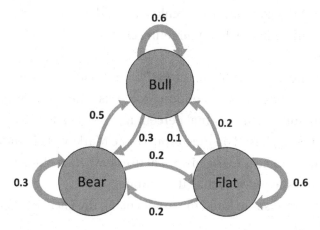

Figure 4-2. *State transition diagram: Markov chain*

This matrix tells us that there is a 60 percent chance of staying in the bull market on a given day, a 30 percent chance of moving to the bear market, and a 10 percent chance of moving to the flat market. Similarly, there is a 30 percent chance of staying in the bear market on a given day, a 20 percent chance of moving to the flat market, and a 50 percent chance of moving to the bull market. Finally, there is a 20 percent chance of moving from the flat market to the bull market, a 20 percent chance of moving to the bear market, and a 60 percent chance of staying in the flat market.

Let's understand how Markov chain models can be used for allocating a budget for different marketing channels/touchpoints.

Problem Statement:

Company X is launching a new Android/iOS expense manager application and wants to create awareness among its target audience. It plans to offer a one-month free trial of the application to people who subscribe through the website. To achieve their marketing goal, they need to identify effective marketing channels that can reach their target audience and generate a high ROI through website sign-ups. While they have data from a previous similar campaign and have been using a heuristic model (last touch attribution) for budget allocation, they want to leverage their ML team's inputs to build a more data-driven solution that can analyze and optimize their marketing channels.

Solution:

Based on the given problem, the ML team suggested using a Markov chain model as it best suits attribution problems. The primary aim of the team is to find out the optimal percentage budget allocation for each marketing channel, based on the output of the Markov chain model.

The team has access to the historical user path data along with the daily marketing spend for each of the channel.

The team proposes to build the model using the ChannelAttribution Python package.

Step 1: Importing the required libraries

```
[In]:
import numpy as np
import pandas as pd
from ChannelAttribution import *
```

Step 2: Reading historical campaign data (user path and daily marketing spend for the channels)

```
[In]:
attribution_data = pd.read_csv('attribution_data.csv')
attribution_data['time'] = pd.to_datetime(attribution_
data['time'])
daily_budget = pd.read_csv('daily_budget.csv')
```

Step 3: Data preprocessing

We create the path order for when users interact with different channels (this is equivalent to SQL window functions).

```
[In]:
attribution_data['path_order'] = attribution_data.sort_
values(['time']).groupby(['cookie']).cumcount() + 1
```

We aggregate the channels a user interacted with into a single row.

```
[In]:
attribution_data_paths = attribution_data.groupby('cookie')
['channel'].agg(lambda x: x.tolist()).reset_index()
attribution_data_paths = attribution_data_paths.
rename(columns={'channel': 'path'})
```

We aggregate conversions and their value by cookie ID.

```
[In]:
attribution_data_conversions = attribution_data.
groupby('cookie', as_index=False).agg({'conversion': 'sum',
'conversion_value': 'sum'})
```

We append path data with that of conversion.

```
[In]:
attribution_data_final = pd.merge(attribution_data_paths,
attribution_data_conversions, how='left', on='cookie')
[In]:
print('Total conversions: {}'.format(sum(attribution_data.
conversion)))
print('Total conversion rate: {}%'.
format(round(sum(attribution_data.conversion) /
len(attribution_data)*100)))
print('Total value of conversions: ${}'.
format(round(sum(attribution_data.conversion_value))))
print('Average conversion value: ${}'.
format(round(sum(attribution_data.conversion_value) /
sum(attribution_data.conversion))))
[Out]:
Total conversions: 19613
Total conversion rate: 3%
Total value of conversions: $122529
Average conversion value: $6
```

We will create a variable path in the specific format required by the Attribution Model package where the ordered channels a user interacts with are delimited by >.

```
[In]:
def listToString(df):
    str1 = ""
    for i in df['path']:
        str1 += i + ' > '
    return str1[:-3]

attribution_data_final['path'] = attribution_data_final.
apply(listToString, axis=1)
```

We remove the user's cookie and grouping by the path to see how many times a specific combination of channels led to a conversion or null outcome.

```
[In]:
attribution_data_final.drop(columns = 'cookie', inplace = True)
attribution_data_final['null'] = np.where(attribution_data_
final['conversion'] == 0,1,0)

attribution_data_final = attribution_data_final.
groupby(['path'], as_index = False).sum()
attribution_data_final.rename(columns={"conversion": "total_
conversions", "null": "total_null", "conversion_value": "total_
conversion_value"}, inplace = True)
```

Step 4: Building the attribution models (heuristic for comparison and Markov models for calculations) and saving the results in a dataframe

```
[In]:
heuristic_models = heuristic_models(attribution_data_
final,"path","total_conversions",var_value="total_
conversion_value")
markov_model = markov_model(attribution_data_final, "path",
"total_conversions", var_value="total_conversion_value")
[Out]:
Number of simulations: 100000 - Convergence reached:
0.85% < 5.00%
Percentage of simulated paths that successfully end before
maximum number of steps (456) is reached: 99.99%
[In]:
all_attribution_models_result = pd.merge(heuristic_
models,markov_model,on="channel_name",how="inner")
```

```
all_attribution_models_conversion_result = all_attribution_
models_result[["channel_name","first_touch_conversions","last_
touch_conversions",\
"linear_touch_conversions","total_conversions"]]
all_attribution_models_conversion_result.columns = ["channel_
name","first_touch","last_touch","linear_touch","markov_model"]
all_attribution_models_conversion_value_result = all_
attribution_models_result[["channel_name","first_touch_
value","last_touch_value",\
"linear_touch_value","total_conversion_value"]]
all_attribution_models_conversion_value_result.columns
= ["channel_name","first_touch","last_touch","linear_
touch","markov_model"]
all_attribution_models_conversion_value_result =
pd.melt(all_attribution_models_conversion_value_result, id_
vars="channel_name")
```

Step 5: Calculating the optimal budget for the marketing channels

```
[In]:
daily_budget_agg = daily_budget.drop(['day', 'impressions'],
axis=1).groupby('channel', as_index=False).sum('cost')
daily_budget_agg.columns = ['channel_name', 'cost']
roas_data = pd.merge(all_attribution_models_conversion_result,
daily_budget_agg)
roas_data['channel_weight'] = (roas_data['markov_model'])/
(sum(roas_data['markov_model']))
roas_data['cost_weight'] = (roas_data['cost'])/(sum(roas_
data['cost']))
roas_data['roas'] = (roas_data['channel_weight'])/(roas_
data['cost_weight'])
```

```python
roas_data['optimal_budget'] = (roas_data['cost'])*(roas_
data['roas'])
roas_data['optimal_budget_percentage'] = (round((roas_
data['optimal_budget'])/(sum(roas_data['optimal_
budget']))),2))*100
roas_data.head()
[Out]:
```

channel_name	markov_model	cost	channel_weight	cost_weight	roas	optimal_budget	optimal_budget_percentage
Facebook	5289.103801	1481.679	0.26967337	0.305076822	0.883952337	1309.733172	27
Instagram	4468.380797	641.159	0.227827502	0.1320143	1.725778963	1106.498714	23
Paid Search	3957.708706	1187.142	0.201790073	0.244431809	0.825547518	980.0417185	20
Online Display	2395.295285	554.937	0.12212794	0.114261236	1.068848407	593.1435285	12
Online Video	3502.511411	991.823	0.178581115	0.204215833	0.874472427	867.3218664	18

From the previous output, the team realized that Instagram and online display channels, while performing well in driving the conversions, were under-budgeted. They need to be given a higher share of the marketing budget, which, in this case is 23 percent and 12 percent for Instagram and online display, respectively.

Decision Intelligence Methodology: AI/ML Models

AI/ML models are based on machine learning algorithms that can learn from data and improve their performance over time. These models are used in a wide range of applications, from image recognition and natural language processing to predictive analytics and decision-making.

Unlike mathematical and probabilistic models, AI models do not require explicit formulas or assumptions about the relationships between variables. Instead, they learn patterns and relationships from large datasets and can identify complex patterns and trends that are difficult or impossible for humans to discern.

Let's understand how AI/ML models can be used for decision intelligence through an HR analytics use case.

Problem Statement:

A multinational corporation with nine distinct areas of operation faces a challenge in selecting suitable candidates for managerial (and below positions) promotions and providing them with timely training. The announcement of final promotions occurs only after evaluations, resulting in delays in transitioning to new roles. Therefore, the company seeks assistance in identifying qualified candidates at specific checkpoints to accelerate the promotion process as a whole.

The company has historical data comprising demographic, educational, work experience, and skill details, along with the promotion details.

Solution:

As the company has access to historical data, it makes sense to build a machine learning model that can find the traits of employees fit for promotion. As the goal is to find the list of employees who can be promoted, it will come under the supervised classification category.

Step 1: Importing the libraries and reading the training and test data

```
[In]:
from sklearn.preprocessing import LabelEncoder, StandardScaler,
RobustScaler
from sklearn.model_selection import train_test_split,
StratifiedKFold
from sklearn.linear_model import LogisticRegression
from sklearn.impute import SimpleImputer
from sklearn.metrics import *
import pandas as pd
import numpy as np

[In]:
training_data = pd.read_csv('hr_train.csv')
test_data = pd.read_csv('hr_test.csv')
```

Step 2: Data preprocessing

As the employee_id column is an ID column, it serves no purpose in the modeling process. So, we delete the employee_id column from the training and test data.

```
[In]:
del training_data['employee_id']
del test_data['employee_id']
categorical_columns = [c for c in training_data.columns if
training_data[c].dtypes=='object']
```

```
numeric_columns = [n for n in training_data.columns if n not in
categorical_columns]
true_categorical_columns = ['department', 'region',
'education', 'gender', 'recruitment_channel','awards_won?',
            'previous_year_rating','length_of_service',
            'no_of_trainings']
true_numeric_columns = [c for c in training_data.columns if c
not in true_categorical_columns]
true_numeric_columns.remove('is_promoted')
```

There are very few employees whose work experience is 30 years or more (0.1 percent). These people can be considered as outliers, and it is a best practice to remove them before we feed data to the model.

```
[In]:
training_data = training_data[training_data.length_of_
service<30]
```

We apply the imputation strategy for the education and previous_ year_rating columns.

```
[In]:
target = training_data.pop('is_promoted')
all_data = pd.concat([training_data, test_data], axis=0)
si_most_frequent = SimpleImputer(strategy= 'most_frequent')
all_data['education'] = si_most_frequent.fit_transform(all_
data.education.values.reshape(-1, 1))
si_mean = SimpleImputer(strategy='mean')
all_data['previous_year_rating'] = si_mean.fit_transform(all_
data.previous_year_rating.values.reshape(-1, 1))
```

We convert categorical features into dummy ones.

```
[In]:
all_data_dummy = pd.get_dummies(all_data)
new_training_data = all_data_dummy.iloc[:len(training_data)]
new_test_data = all_data_dummy.iloc[len(training_data):]
```

We apply StandardScaler for standardizing the features in training and test data.

```
ss = StandardScaler()
new_training_data = ss.fit_transform(new_training_data)
new_test_data = ss.transform(new_test_data)
```

Step 3: Model training (logistic regression) and evaluation

We perform stratified K-fold cross validation on the data and training model and evaluate the performance. As the purpose of this activity is to show how AI/ML model outputs can be used for decision intelligence, we will keep the modeling process pretty simple and not train multiple algorithms to see which ones perform the best. Instead, we will train a logistic regression model with default/minimal changes to the parameters and will see how the results can be interpreted.

```
[In]:
scores = []
probs = np.zeros(len(new_training_data))
y_le = target.values
folds = StratifiedKFold(n_splits=5, shuffle=True, random_
state=42)

for fold_, (train_ind, val_ind) in enumerate(folds.split(new_
training_data, y_le)):
    print('fold:', fold_)
    X_tr, X_test = new_training_data[train_ind], new_training_
    data[val_ind]
    y_tr, y_test = y_le[train_ind], y_le[val_ind]
```

```
clf = LogisticRegression(max_iter=200, random_state=2020)
clf.fit(X_tr, y_tr)
probs[val_ind]= clf.predict_proba(X_test)[:, 1]
y = clf.predict_proba(X_tr)[:,1]
print('train:',roc_auc_score(y_tr, y),'val :' , roc_auc_
score(y_test, (probs[val_ind])))
print(20 * '-')
scores.append(roc_auc_score(y_test, probs[val_ind]))
print('log reg  roc_auc=  ', round(np.mean(scores)*100,2))
probs_rnd = np.where(probs > 0.5, 1, 0)
print('F1 Score= ', round(f1_score(target, probs_rnd)*100,2))
print('Recall Score= ', round(recall_score(target, probs_
rnd)*100,2))
[Out]:
fold: 0
train: 0.7931202718718571 val : 0.7819824102253672
--------------------
fold: 1
train: 0.7897501709069539 val : 0.79278434937156
--------------------
fold: 2
train: 0.7868061847019592 val : 0.804318671120831
--------------------
fold: 3
train: 0.7930340720726938 val : 0.7815370354855481
--------------------
fold: 4
train: 0.7943494182608088 val : 0.7806564317616108
--------------------
log reg  roc_auc=  78.83
F1 Score=  44.66
Recall Score=  29.24
```

As we can see, the model performance is satisfactory, and with additional techniques, it can be improved. Now, the HR team can leverage the predictions made on the employees through the ML model and create a list of potential promotion candidates based on their probability scores.

Conclusion

In this chapter, we went through practical real-world use-case problems and how they can be solved through different decision intelligence methodologies. We also saw how the output from the different models can be used for interpretation and how one can build a decision intelligence workflow from it. We have specifically focused on some of the most used mathematical, probabilistic, and AI/ML models, as covering all methodologies in a single chapter/book is impossible.

CHAPTER 5

Augmenting Decision Intelligence Results into the Business Workflow

In today's rapidly changing business environment, organizations are under increasing pressure to make informed decisions that drive growth and competitiveness. Decision intelligence (DI) is a powerful tool that can help organizations make better decisions by leveraging the power of artificial intelligence (AI) and data analytics. However, to fully realize the benefits of DI, it is essential to integrate the outputs of DI models into existing business workflows.

In this chapter, we will discuss how to effectively integrate DI into business workflows through the development of user-friendly applications that connect AI predictions to existing business tools, the importance of collaboration between DI experts and business teams, and some of the technical challenges involved in integrating DI into existing business systems. By successfully integrating DI into business workflows, organizations can leverage the power of AI to drive better decision-making and improve business outcomes.

© Akshay Kulkarni, Adarsha Shivananda, Avinash Manure 2023
A. Kulkarni et al., *Introduction to Prescriptive AI*,
https://doi.org/10.1007/978-1-4842-9568-7_5

Challenges

Integrating decision intelligence with business workflows can be a complex process and comes with several challenges. Here are some of the main challenges organizations may face:

- **Data quality and availability:** One of the primary challenges is ensuring that the data used by the DI process is of high quality and readily available. Data quality issues, such as missing or incorrect data, can impact the accuracy and usefulness of the insights generated by the DI process.

- **Model selection and validation:** Another challenge is selecting the appropriate machine learning models for the specific business problem and validating their accuracy and reliability. This requires expertise in both data science and the specific domain being analyzed.

- **Integration with existing systems:** Integrating the AI outputs with existing systems can be challenging. It requires a thorough understanding of the organization's IT infrastructure and security protocols to ensure that the app can be integrated smoothly and securely.

- **User adoption and training:** Even the best-designed app may not be effective if users are not willing or able to adopt it. Ensuring user adoption and providing adequate training is critical to the success of the app.

- **Ongoing maintenance and updates:** Like all software applications, even this require ongoing maintenance and updates to ensure that it remains functional and up-to-date with the latest technology and data. This requires resources and expertise to ensure that the app remains effective over time.

Especially integrating decision intelligence with existing systems can be challenging for several reasons.

- **Different data formats:** Existing systems may store data in different formats, making it difficult to integrate with the DI system. For example, one system may use a CSV file format, while another system may use a database format. Converting data between formats can be time-consuming and error prone.

- **Legacy systems:** Some organizations may still be using legacy systems that are not designed to integrate with modern software applications. These systems may have limited capabilities and may not support modern integration methods, such as REST APIs.

- **Security considerations:** Existing systems may have strict security protocols in place to protect sensitive data. Integrating with these systems may require additional security measures, such as authentication and encryption, to ensure that data is protected.

- **Scalability:** Existing systems may not be designed to handle the increased workload that comes with integrating with a DI system. This can lead to performance issues and may require additional hardware or infrastructure to support the integration.

- **Complexity:** Existing systems may be complex and difficult to understand, especially if they have been in use for many years. This can make it challenging to identify the data required for the DI system and to design an integration strategy that works effectively.

Overall, integrating DI with existing systems requires a thorough understanding of the existing IT infrastructure, as well as expertise in data integration and system architecture. Addressing these challenges is critical to ensure that the integration is successful and that the DI system can provide accurate and reliable insights to support decision-making.

Workflow

Let's understand the steps involved to ensure DI outputs are properly ingested into business workflows and overcome the previous challenges.

1. **Clearly define the business problem:** Before starting the DI process, it is important to clearly define the business problem that you are trying to solve. This will help you to identify the relevant data sources and develop machine learning models that are aligned with your business objectives.

2. **Involve key stakeholders:** Involve key stakeholders in the DI process, including business users, data scientists, and IT professionals. This will help to ensure that the DI outputs are aligned with business needs and can be integrated into existing business workflows.

3. **Integrations:** Develop user-friendly interfaces that can be easily integrated into existing business workflows. This may involve developing APIs that can be called from within business applications

or integrating the models into existing business processes. Business workflow could be in CRM, marketing application, apps, etc.

4. **Ensure data quality:** Ensure that the data used in the DI process is accurate, complete, and relevant. This may require data cleansing, normalization, and transformation.

5. **Monitor performance:** Monitor the performance of the DI process and evaluate its impact on business outcomes. This will help you to identify areas for improvement and make adjustments as necessary.

6. **Provide training and support:** Provide training and support to business users on how to use the DI outputs in their daily work. This will help to ensure that the DI outputs are properly ingested into business workflows and that the organization can derive maximum value from the DI process.

As shown in Figure 5-1, organizations can ensure that DI outputs are properly ingested into business workflows, leading to better decisions, improved operations, and a competitive advantage.

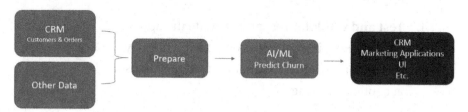

Figure 5-1. *Workflow*

Decision Intelligence Apps

One of the ways businesses can use predictions is through apps. So, let's explore how to build/integrate apps with DI and why this is important.

Integrating a DI framework through apps involves developing applications that can leverage the insights generated by the DI process. Here are some steps to consider when integrating DI through apps:

1. **Define the business problem:** Clearly define the business problem you are trying to solve using DI. This will help you to identify the relevant data sources and determine the appropriate machine learning models to use.

2. **Develop the DI process:** Develop the DI process to generate insights that are relevant to the business problem. This may involve data cleansing, feature engineering, model training, and validation.

3. **Develop the app:** Develop an app that can leverage the insights generated by the DI process. This may involve developing APIs that can be called from within the app or integrating the models into the app's functionality.

4. **Test and validate:** Test and validate the app to ensure that it is working as expected. This may involve unit testing, integration testing, and user acceptance testing.

5. **Deploy the app:** Deploy the app to a production environment where it can be used by business users. This may involve working with IT professionals to ensure that the app is secure, scalable, and reliable.

6. **Monitor performance:** Monitor the performance
    of the app and the DI process to identify areas for
    improvement and make adjustments as necessary.

How and Why?

Developing an app to integrate DI involves creating a software application
that leverages the insights generated by the DI process. Here are some
details on how and why to develop an app for DI:

1. **Identify the use case:** The first step is to identify
    the use case for the app. What business problem are
    you trying to solve with DI? What insights do you
    need to generate to address this problem? Once you
    have a clear understanding of the use case, you can
    determine the functionality required for the app.

2. **Define the user interface:** The next step is to
    define the user interface for the app. This involves
    identifying the user personas and designing an
    interface that is easy to use and understand. The
    interface should provide users with access to the
    insights generated by the DI process in a clear and
    intuitive manner.

3. **Develop the app functionality:** Once you have
    defined the user interface, the next step is to develop
    the functionality for the app. This may involve
    developing APIs that can be called from within
    the app to access the insights generated by the DI
    process. It may also involve integrating the machine
    learning models into the app's functionality to
    generate predictions or recommendations based on
    the user's input.

4. **Test and validate the app:** Once the app has been developed, it is important to test and validate it to ensure that it is working as expected. This may involve unit testing, integration testing, and user acceptance testing. The testing phase helps to identify any bugs or issues that need to be addressed before the app is deployed.

5. **Deploy the app:** Once the app has been tested and validated, it is ready to be deployed to a production environment. This may involve working with IT professionals to ensure that the app is secure, scalable, and reliable.

By developing an app to integrate DI, organizations can provide business users with easy access to the insights generated by the DI process. The app can help to streamline workflows, improve decision-making, and drive business value. However, it is important to ensure that the app is user-friendly, reliable, and secure to maximize adoption and ensure that the organization can derive maximum value from the DI process.

User-Friendly Interfaces

Designing user-friendly interfaces is essential when integrating decision intelligence outputs into business workflows. Because the users are nontechnical, it's important to develop interfaces that are easy to use and take actions.

Here are some reasons why a user-friendly interface matters:

- **Facilitates adoption:** User-friendly interfaces make it easier for business users to adopt and use the DI outputs in their daily work. If the interfaces are difficult to use, it may discourage adoption, which could undermine the effectiveness of the DI process.

- **Enhances usability:** User-friendly interfaces enhance the usability of the DI outputs. By making the interfaces easy to use and understand, business users can quickly access the insights they need and make better decisions.

- **Improves productivity:** User-friendly interfaces can improve productivity by reducing the time and effort required to access the DI outputs. This can help to streamline business workflows, increase efficiency, and reduce costs.

- **Supports collaboration:** User-friendly interfaces can support collaboration between business users and data scientists. By providing a common interface for accessing DI outputs, business users and data scientists can work together more effectively and share insights more easily.

- **Increases engagement:** User-friendly interfaces can increase engagement with the DI outputs. By providing a visually appealing and intuitive interface, business users are more likely to engage with the insights provided by the DI process.

In summary, designing user-friendly interfaces is critical when integrating DI outputs into business workflows. It can facilitate adoption, enhance usability, improve productivity, support collaboration, and increase engagement. By prioritizing user experience, organizations can maximize the value derived from the DI process and achieve better business outcomes.

Augmenting AI Predictions to Business Workflow

Connecting AI predictions to business tools involves integrating the AI system with existing business systems, such as enterprise resource planning (ERP) systems, customer relationship management (CRM) systems, and business intelligence (BI) tools. Here are the technical steps to connect AI predictions to business tools:

1. **Collect and process data:** Collect and process data required for the AI model to make predictions. This data can be sourced from internal business systems or external sources.

2. **Build and train the AI model:** Develop the AI model that will generate predictions based on the data. Train the model using historical data to ensure accurate predictions.

3. **Deploy the AI model:** Deploy the trained AI model to an environment where it can be accessed by business tools. This can be done using application programming interfaces (APIs), web services, or other integration tools.

4. **Connect to business tools:** Connect the AI model to the relevant business tools, such as ERP systems, CRM systems, or BI tools. This can be done through APIs, custom integrations, or prebuilt connectors.

5. **Test and validate:** Test the integration to ensure that AI predictions are properly integrated into the business tools. Validate that the predictions align with the business goals and objectives.

6. **Monitor and maintain:** Monitor the AI system to ensure that it continues to provide accurate predictions. Maintenance and updates may be required to address changes in the business environment or to improve the accuracy of the predictions.

As shown in Figure 5-2, in summary, connecting AI predictions to business tools involves data collection and processing, building and training an AI model, deploying the model, connecting to business tools, testing and validating the integration, and ongoing monitoring and maintenance. By following these technical steps, organizations can ensure that AI predictions are effectively integrated into their business processes and support informed decision-making

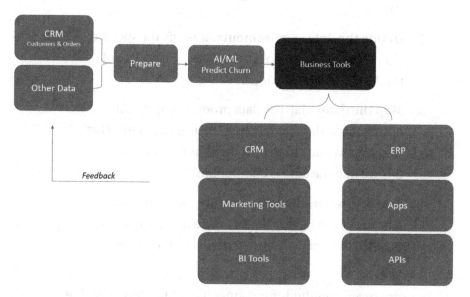

Figure 5-2. *Augmenting predictions with business tools*

Connect to Business Tools

Connecting AI predictions to business tools means connecting the AI model to the relevant business tools, such as ERP systems, CRM systems, or BI tools. This step is crucial because it enables the AI model to share its predictions with other systems that are critical to business operations.

Connecting the AI model to business tools involves integrating the AI model with the business tool's API. APIs provide a standard interface for different systems to communicate with each other, enabling data to be exchanged seamlessly between systems. The integration process typically involves the following steps:

1. **Identify the appropriate API:** Determine which API the business tool supports and what data format it uses.

2. **Define the data requirements:** Identify the specific data that the business tool requires to make use of the AI model's predictions.

3. **Map the data:** Map the data produced by the AI model to the data required by the business tool. This may involve data transformation or conversion to ensure that the data is compatible.

4. **Test the integration:** Test the integration to ensure that data is being transferred correctly and that the business tool can use the AI model's predictions effectively.

There are several methods for connecting AI models to business tools, including custom integration, prebuilt connectors, and third-party integration platforms. The choice of method depends on the specific requirements of the organization and the systems being integrated.

In summary, connecting AI predictions to business tools involves integrating the AI model with the business tool's API, mapping data between the two systems, and testing the integration to ensure that it works effectively. By connecting AI models to business tools, organizations can leverage AI predictions to support informed decision-making and improve business outcomes.

Map the Data

Mapping the data in the context of connecting AI predictions to business tools refers to the process of identifying the relevant data produced by the AI model and aligning it with the data required by the business tool. The data produced by the AI model may be in a format that is different from what the business tool requires, so mapping is necessary to ensure that the data can be effectively shared between the two systems.

For example, as shown in Figure 5-3, let's say that an organization has an AI model that predicts customer churn. The AI model produces a list of customers who are at high risk of churning, along with a probability score for each customer. To integrate this prediction into the organization's CRM system, the organization would need to map the data from the AI model to the data required by the CRM system.

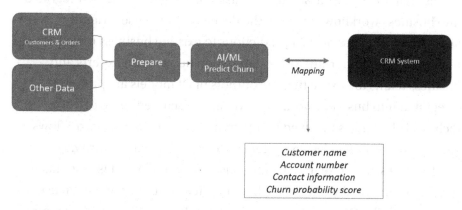

Figure 5-3. Example AI model that predicts customer churn

Mapping the data would involve identifying the specific fields in the CRM system that correspond to the data produced by the AI model. For example, the CRM system may require the customer name, account number, and contact information, in addition to the churn probability score. The organization would need to ensure that the data produced by the AI model is aligned with the data required by the CRM system and may need to perform data transformations or conversions to ensure compatibility.

Once the data mapping is complete, the organization can then connect the AI model to the CRM system, using the CRM system's API or other integration methods, to share the AI predictions with the CRM system. This enables the organization to leverage the AI predictions to inform its customer retention strategies, and to take targeted actions to prevent customer churn.

In summary, mapping the data involves aligning the data produced by the AI model with the data required by the business tool to ensure that the data can be effectively shared and used to support business operations.

Conclusion

In this chapter, we discussed how decision intelligence can be integrated into business workflows through the development of user-friendly applications that connect AI predictions to existing business tools. We highlighted the importance of collaboration between DI experts and business teams to ensure that the outputs of DI models are properly integrated into business workflows. We also discussed some of the technical challenges involved in integrating DI into business workflows, such as connecting to legacy systems and ensuring data security.

To successfully integrate DI into business workflows, it is essential to develop user-friendly interfaces that make it easy for nontechnical users to interact with the DI models and understand the outputs. The development of custom applications that connect AI predictions to existing business tools is an effective way to achieve this. To increase user adoption, it is important to involve business teams in the development process and to provide training and support to help users understand how to use the applications effectively.

Overall, integrating DI into business workflows requires a combination of technical expertise, collaboration between DI experts and business teams, and a focus on developing user-friendly interfaces that make it easy for nontechnical users to interact with the outputs of DI models. By successfully integrating DI into business workflows, organizations can leverage the power of AI to support informed decision-making and improve business outcomes.

CHAPTER 6

Actions, Biases, and Human-in-the-Loop

Ethical AI refers to the development and use of artificial intelligence (AI) in a way that aligns with ethical principles and values. This includes ensuring that AI systems are transparent, accountable, and unbiased, and that they do not cause harm to individuals or society. In this chapter, we will uncover the ethical aspects of AI such as bias and fairness, responsible AI, and human in the loop.

Key Ethical Considerations in AI

Several key ethical considerations must be taken into account when developing AI systems. This includes ensuring that AI systems are designed and implemented in ways that respect individual privacy and autonomy, that they are not used to create or perpetuate unjust power structures, and that they do not contribute to social or economic inequality.

To promote ethical AI, organizations and developers should follow several best practices, such as involving diverse stakeholders in the design and development process, conducting thorough testing and evaluation of AI systems to identify potential biases and unintended consequences, and implementing clear guidelines and standards for ethical AI across the industry.

© Akshay Kulkarni, Adarsha Shivananda, Avinash Manure 2023
A. Kulkarni et al., *Introduction to Prescriptive AI*,
https://doi.org/10.1007/978-1-4842-9568-7_6

- **Bias and fairness:** AI systems can perpetuate and amplify biases, particularly if the data used to train them is biased. Ethical AI requires addressing and mitigating these biases to ensure fairness and equity.

- **Privacy and security:** AI systems often deal with sensitive personal data, such as medical records or financial information. Ethical AI requires ensuring that this data is handled securely and in accordance with relevant privacy laws and regulations.

- **Transparency and explainability:** AI systems can be difficult to understand and interpret, making it challenging to determine how decisions are made. Ethical AI requires transparency and explainability so that individuals can understand how and why decisions are being made.

- **Accountability:** Ethical AI requires accountability so that individuals or organizations responsible for developing and deploying AI systems can be held responsible for any negative outcomes.

- **Human values and rights:** AI systems should be designed and used in a way that respects human values and rights, including autonomy, privacy, and dignity.

Addressing these ethical considerations requires collaboration and engagement across a range of stakeholders, including technologists, policymakers, ethicists, and civil society groups. It also requires ongoing evaluation and adaptation as AI technologies continue to evolve.

Actions, Biases, and Human-in-the-Loop

Actions, biases, and human in the loop are all important considerations when implementing decision intelligence (DI) to ensure that the decisions made are accurate, ethical, and aligned with business goals. Here's a brief explanation of each:

- **Actions:** In DI, the actions taken based on the insights generated by the system are critical. It is important to ensure that the actions are aligned with business goals and are ethical. DI should be used to augment human decision-making, not replace it entirely. Therefore, it's essential to ensure that the actions taken are consistent with organizational values and principles.

- **Biases:** One of the most significant challenges in AI and DI is the potential for biases in the data and models used. Biases can result from various factors, including the collection of data, selection of algorithms, or training data. Therefore, it's crucial to identify and mitigate any biases that may exist in the data or models used. This can be achieved by using diverse data sources, testing and evaluating models, and continuously monitoring and updating the models as needed.

- **Human in the loop:** The concept of human in the loop refers to the involvement of humans in the decision-making process. Despite the advances in AI and DI, human expertise and judgment are still crucial in many decision-making processes. Therefore, it's essential to have a human in the loop to monitor and evaluate the decisions made by the system. This can be achieved by providing humans with visibility into the decision-making process, allowing them to intervene when necessary.

In summary, when implementing DI, it's essential to ensure that the actions taken are consistent with business goals, identify and mitigate biases, and involve humans in the decision-making process. By considering these factors, businesses can ensure that the decisions made using DI are accurate, ethical, and aligned with organizational value.

Cognitive Biases

Cognitive biases can also play a significant role in the development and implementation of artificial intelligence (AI) and machine learning (ML) algorithms. These biases can arise from a variety of sources, including the data used to train the algorithms, the assumptions and biases of the developers and users, and the design of the algorithms themselves.

For example, if the data used to train an AI system is biased, such as containing gender or racial biases, then the system may produce biased results. This is because the AI system learns from the data it is trained on, and if the data contains biased patterns, those patterns can be replicated in the output of the system.

Another example is the bias in the selection of features used to train the AI system. If the selected features are not representative of the problem domain, the AI system may produce inaccurate or biased results.

In addition, cognitive biases of the developers and users can also influence the design and implementation of AI and ML systems. For instance, confirmation bias may lead to overlooking important factors in the development process, while anchoring bias may result in over-reliance on certain data or features.

To address these potential biases in AI and ML, it is essential to promote diversity and inclusivity in data collection and analysis. This can include using diverse datasets, including diverse perspectives in the development team, and testing and validating the algorithms with diverse groups of users.

Furthermore, there are various techniques for detecting and mitigating cognitive biases in AI and ML, such as fairness testing and bias mitigation methods. These techniques help to ensure that AI and ML algorithms are free from cognitive biases and produce results that are accurate and unbiased.

Why Is Detecting Bias Important?

It is important to detect cognitive biases because they can lead to inaccurate and suboptimal decisions. When individuals or organizations make decisions based on biased thinking, they may not be considering all of the available information or may be giving undue weight to certain factors, which can result in poor outcomes.

For example, a hiring manager who is affected by confirmation bias may select only those job candidates who share their own background and beliefs, rather than objectively evaluating the qualifications of all candidates. This can result in a less diverse and less qualified workforce, which can ultimately harm the organization's performance.

Similarly, an investor who is affected by availability bias may make decisions based on the most recent or easily accessible information, rather than taking a more comprehensive and objective view of the market. This can lead to poor investment decisions and financial losses.

By detecting and addressing cognitive biases, individuals and organizations can make more informed and accurate decisions, which can lead to better outcomes. It can also help to promote fairness, diversity, and inclusivity in decision-making, which are important values in many contexts.

Overall, detecting cognitive biases is important because it allows individuals and organizations to make decisions based on a more accurate and comprehensive understanding of the situation, which can lead to better outcomes and more successful outcomes.

Types

Several types of biases can occur in AI and ML algorithms. Here are some of the most common types:

- **Sampling bias:** This occurs when the data used to train an algorithm is not representative of the population it is intended to serve. For example, if an algorithm is trained on data that over-represents a particular demographic group, it may produce biased results when applied to the broader population.

- **Confirmation bias:** This occurs when an algorithm is designed to seek out and reinforce pre-existing beliefs or assumptions. For example, an algorithm used to identify potential job candidates may prioritize certain skills or experiences that are associated with a particular demographic group.

- **Selection bias:** This occurs when an algorithm selectively considers certain features or data points while ignoring others. For example, an algorithm used to evaluate creditworthiness may consider factors such as income and credit history but not consider other important factors such as education level or job stability.

- **Automation bias:** This occurs when humans place too much trust in the decisions made by an algorithm, without adequately questioning or scrutinizing the results. For example, a medical diagnosis algorithm may identify a particular condition, but a doctor should still use their clinical judgment and evaluate the result in the context of the patient's other symptoms and medical history.

- **Group attribution bias:** This occurs when an algorithm attributes certain characteristics or behaviors to an entire group, based on the actions of a few individuals within that group. For example, an algorithm used to predict crime rates may unfairly attribute certain criminal behaviors to specific demographic groups.

- **Algorithmic bias:** This occurs when the design or implementation of an algorithm itself is inherently biased. For example, an algorithm may use certain features or decision-making processes that unfairly advantage or disadvantage certain groups.

Detecting and mitigating these biases requires careful attention to the design and implementation of algorithms, as well as ongoing monitoring and evaluation to ensure that the algorithms are producing accurate and fair results.

What Happens If Bias Is Ignored?

Bias can have a significant impact on an organization in several ways.

- **Decreased performance:** Biased decision-making can lead to poor outcomes and decreased performance. When decisions are made based on flawed assumptions or incomplete information, organizations may miss opportunities or make suboptimal choices that negatively impact their bottom line.

- **Damage to reputation:** Biased decision-making can also damage an organization's reputation. If an organization is perceived as being unfair or discriminatory, it can lose the trust and confidence of customers, employees, and other stakeholders, which can have long-term consequences.

- **Legal and regulatory risks:** Biased decision-making can also expose organizations to legal and regulatory risks. Discriminatory practices can lead to lawsuits, fines, and other legal penalties, which can be costly and damaging to an organization's reputation.

- **Reduced diversity and inclusion:** Biased decision-making can also lead to reduced diversity and inclusion within an organization. If decisions are made based on preconceived notions or stereotypes, it can limit opportunities for individuals from diverse backgrounds and create a less inclusive workplace culture.

- **Inefficient use of resources:** Biased decision-making can also result in an inefficient use of resources. If decisions are made based on flawed assumptions or incomplete information, organizations may invest

resources in projects or initiatives that are unlikely
to succeed, which can waste time, money, and other
resources.

Overall, bias can have a significant impact on an organization's
performance, reputation, and ability to attract and retain talented
employees. By detecting and addressing biases in decision-making,
organizations can promote fairness, diversity, and inclusion, and improve
their overall performance and success.

Bias Detection

Detecting bias in AI and ML algorithms can be challenging, but there are
several approaches that can be taken to identify and mitigate bias.

- **Data analysis:** One way to detect bias is to conduct a
  comprehensive analysis of the data used to train the
  algorithm. This analysis should include examining the
  data for any patterns or biases that may be present, as
  well as identifying any gaps or omissions in the data.

- **Auditing:** Another way to detect bias is to conduct
  an audit of the algorithm to evaluate its performance
  and accuracy. This audit should include evaluating
  the algorithm's outputs and assessing its impact on
  different groups.

- **Testing:** Testing the algorithm on different inputs
  and scenarios can help identify any biases or
  inconsistencies in its outputs. This testing should
  include both positive and negative examples to ensure
  that the algorithm is producing accurate results across
  different scenarios.

- **Expert review:** Having experts in the relevant field review the algorithm can provide valuable insights into potential biases and how to mitigate them. These experts can provide feedback on the algorithm's design and implementation and suggest ways to improve its performance and accuracy.

- **Bias detection tools:** There are also a growing number of tools available that are designed specifically to detect and mitigate bias in AI and ML algorithms. These tools use a variety of techniques, including data analysis and machine learning, to identify potential biases and suggest ways to mitigate them.

Overall, detecting bias in AI and ML algorithms requires a combination of data analysis, testing, expert review, and use of bias detection tools. By employing these approaches, organizations can ensure that their algorithms are producing accurate and fair results and promote fairness, diversity, and inclusivity in their decision-making processes.

What Do Bias Tools Do?

Bias detection tools are software applications that use a variety of techniques to identify and mitigate bias in AI and ML algorithms. These tools typically work by analyzing the data used to train the algorithm, evaluating the algorithm's outputs, and providing feedback on potential biases and ways to mitigate them.

These are some common techniques used by bias detection tools:

- **Data analysis:** Many bias detection tools use data analysis techniques to identify patterns and biases in the data used to train the algorithm. These tools may look for correlations between different variables and evaluate the quality and diversity of the data.

- **Machine learning:** Some bias detection tools use machine learning algorithms to train models that can detect bias in other algorithms. These models are trained on large datasets and can identify patterns and biases that may be difficult to detect manually.

- **Statistical analysis:** Statistical analysis techniques can also be used to detect biases in AI and ML algorithms. These techniques include regression analysis, hypothesis testing, and other statistical methods that can identify patterns and relationships in the data.

- **Fairness metrics:** Some bias detection tools use fairness metrics to evaluate the algorithm's outputs and identify potential biases. Fairness metrics measure the impact of the algorithm on different groups and can help identify where bias may be present.

- **Human review:** Finally, many bias detection tools incorporate human review into the process. This may involve having experts in the relevant field review the algorithm's outputs and provide feedback on potential biases and ways to mitigate them.

Overall, bias detection tools use a variety of techniques to identify and mitigate bias in AI and ML algorithms. These tools can help organizations ensure that their algorithms are producing fair and accurate results and promote fairness, diversity, and inclusivity in their decision-making processes.

Incorporation of Feedback Through Human Intervention

Incorporating feedback through human intervention in AI and ML algorithms can be done in several ways, depending on the specific algorithm and the nature of the feedback. Here are some common approaches:

- **User feedback:** One way to incorporate feedback is to collect feedback from users of the algorithm. This can be done through surveys, user testing, or other methods. User feedback can provide valuable insights into how the algorithm is performing and where improvements may be needed.

- **Expert review:** Another way to incorporate feedback is to have experts in the relevant field review the algorithm and provide feedback. These experts can provide feedback on the algorithm's design, performance, and accuracy, as well as suggest ways to improve its performance.

- **Human-in-the-loop systems:** Some AI and ML algorithms incorporate a "human-in-the-loop" system, where human intervention is built into the algorithm itself. For example, the algorithm may flag certain outputs for human review or require human intervention for certain decisions.

- **Continuous monitoring:** Finally, continuous monitoring of the algorithm's outputs can also provide valuable feedback. By tracking the algorithm's performance over time, organizations can identify areas where the algorithm may be biased or where improvements are needed.

Incorporating feedback through human intervention can be useful in several ways.

- **Improving accuracy:** By incorporating feedback from users and experts, organizations can identify areas where the algorithm may be inaccurate or biased and make improvements to increase its accuracy.

- **Enhancing fairness:** Feedback from users and experts can also help organizations identify areas where the algorithm may be unfair or discriminatory and make changes to promote fairness and inclusivity.

- **Increasing trust:** Incorporating feedback from users and experts can help increase trust in the algorithm and the organization using it. When users feel that their feedback is being listened to and incorporated into the algorithm, they may be more likely to trust its outputs.

Overall, incorporating feedback through human intervention can help organizations improve the accuracy, fairness, and trustworthiness of their AI and ML algorithms, and promote fairness, diversity, and inclusivity in their decision-making processes.

Human-in-the-loop (HITL) is a design pattern for AI and ML systems where a human is involved in the decision-making process alongside the algorithm. HITL systems are designed to combine the strengths of both humans and machines, with the human providing the judgment and expertise and the machine handling the more tedious and repetitive tasks.

In HITL systems, the algorithm provides a recommendation or output, which is then reviewed or modified by a human operator. The human operator can either approve the algorithm's recommendation or modify it based on their own judgment and expertise. The modified output is then used to train the algorithm, improving its performance over time.

HITL systems can be used in a wide range of applications, including image recognition, natural language processing, and decision-making systems. For example, in an image recognition system, the algorithm may identify objects in an image, which are then reviewed by a human operator to ensure their accuracy. In a decision-making system, the algorithm may provide a recommendation, which is then reviewed by a human operator to ensure its fairness and ethicality.

HITL systems can be useful in several ways.

- **Improved accuracy:** By combining the strengths of both humans and machines, HITL systems can improve the accuracy of the algorithm's outputs. Humans can provide judgment and expertise, which can help the algorithm make better decisions.

- **Increased transparency:** HITL systems can increase transparency in the decision-making process. By involving a human operator, the decision-making process becomes more explainable, making it easier to understand how the algorithm arrived at its recommendation.

- **Better adaptability:** HITL systems are more adaptable than fully automated systems. When faced with new or unexpected situations, human operators can provide judgment and expertise that the algorithm may not possess, allowing the system to handle a wider range of scenarios.

Overall, HITL systems are a powerful tool for improving the accuracy, transparency, and adaptability of AI and ML systems and can help organizations ensure that their algorithms are making fair and ethical decisions.

How to Build HITL Systems?

Building a human-in-the-loop system involves designing an AI or ML system that incorporates human input and feedback into its decision-making process. Here are some steps to consider when building a HITL system:

1. **Identify the problem:** The first step in building an HITL system is to identify the problem you want to solve. This could be anything from image recognition to natural language processing or decision-making.

2. **Choose the right algorithm:** Once you have identified the problem, you need to choose the right AI or ML algorithm to solve it. Consider the strengths and weaknesses of different algorithms, and choose one that is well-suited to your specific problem.

3. **Define the human-in-the-loop process:** Next, you need to define the process for incorporating human input and feedback into the algorithm. This could involve defining the types of decisions that require human review, setting up a user interface for human input, or defining the criteria for accepting or rejecting human input.

4. **Develop the user interface:** Depending on the specific HITL process you have defined, you may need to develop a user interface for humans to interact with the algorithm. This could involve building a web application or mobile app or integrating with an existing software tool.

5. **Collect and incorporate human feedback:** Once your HITL system is up and running, you can begin collecting human feedback and incorporating it into the algorithm. This could involve collecting feedback through surveys or user testing or providing human operators with specific criteria for modifying the algorithm's output.

6. **Continuously monitor and refine the system:** As you collect feedback and incorporate it into the algorithm, it's important to continuously monitor the system's performance and refine it over time. This could involve updating the algorithm to address common errors or biases or modifying the HITL process to improve efficiency or accuracy.

Building an HITL system can be a complex and iterative process, but it can be a powerful tool for improving the accuracy, transparency, and fairness of AI and ML systems. By incorporating human input and feedback into the decision-making process, organizations can ensure that their algorithms are making ethical and fair decisions that align with their values and goals.

Let's understand this more through an example.

Example: Customer Churn

Churn prediction is a common problem in industries such as telecommunications, where customers may cancel their services after a certain period of time. In this context, an action could be to proactively reach out to customers who are at risk of churning in order to retain their business. However, there may be biases in the algorithm that determines which customers are at risk of churning, such as an overemphasis on a certain demographic or a failure to consider certain factors that may influence customer loyalty.

To address these biases, a human-in-the-loop system could be implemented. In this system, the algorithm would provide a recommendation for which customers are at risk of churning, but this recommendation would be reviewed by a human operator before any action is taken. The human operator could review the algorithm's output, assess its fairness and accuracy, and modify the recommendation based on their judgment and expertise.

For example, the algorithm may identify a certain demographic as being at a higher risk of churning, but the human operator may notice that this demographic is actually more loyal than the algorithm suggests. The human operator could modify the recommendation to focus on a different demographic or to take additional factors into account that the algorithm did not consider.

By incorporating human input and feedback into the churn prediction process, the HITL system can improve the accuracy, fairness, and ethicality of the algorithm's recommendations. This can lead to better outcomes for the organization, such as increased customer retention and higher revenue, while avoiding biases that could harm certain groups of customers or lead to unfair decision-making.

Conclusion

In this chapter, we discussed how cognitive biases are a prevalent issue in AI and ML and can have significant negative impacts on organizations and individuals. To address this problem, human-in-the-loop systems can be implemented to incorporate human input and feedback into the decision-making process of algorithms. HITL systems can help detect and mitigate biases, improve the accuracy and fairness of AI and ML systems, and ensure that they align with the values and goals of the organization. Building an HITL system involves identifying the problem, choosing the right algorithm, defining the HITL process, developing the user interface, collecting and incorporating human feedback, and continuously monitoring and refining the system. By using HITL systems, organizations can make more informed and ethical decisions that benefit both the organization and its stakeholders.

CHAPTER 7

Case Studies

This chapter will explore how decision intelligence can be integrated into organizations through a series of case studies. We will delve into various methodologies and examine how they can produce actionable insights for decision-makers, both independently and when used in conjunction with other methods.

While we have already covered many methodologies in prior chapters and demonstrated how they can be applied to decision intelligence, it is uncommon to find a single methodology that can effectively solve a decision intelligence problem. The problems are often so intricate that employing a combination of methodologies is necessary to achieve the desired outcome.

Let's explore how AI/ML along with other advanced techniques can help in decision-making through two case studies: one classification and another regression.

Case Study 1: Telecom Customer Churn Management

Problem Statement: A telecom company is experiencing high customer churn rates and is looking for innovative AI solutions to reduce churn. The company wants to leverage machine learning along with other techniques to do the following:

© Akshay Kulkarni, Adarsha Shivananda, Avinash Manure 2023
A. Kulkarni et al., *Introduction to Prescriptive AI*,
https://doi.org/10.1007/978-1-4842-9568-7_7

- Monitor customer behavior and identify early warning signs of churn

- Identify the factors that lead customers to churn

- Create personalized solutions for retaining the customers and run simulations

- Update the solutions based on the simulation output

Solution:

The AI team proposes a hybrid solution to solve the given problem. The plan is to first build a machine learning model that takes historical customer details such as account information, demographics, and services used, along with churn status. The model output will be used to find potential customers who are likely to churn in the coming time. The team also proposes a counterfactual generation engine that can provide various scenarios wherein the predicted outcomes change. For example, if a customer is likely to churn and their churn probability is 90 percent, the counterfactual engine can generate change in input features that might bring the churn probability for the given customer to maybe 40 percent. This change in the input parameters can be used as part of their retention strategy. A third component of a what-if analysis solution can help decision-makers tweak the counterfactual suggestions based on certain constraints and check whether those changes can still retain customers.

The AI team suggests following benefits of using the proposed hybrid solution:

- **Improved accuracy:** A machine learning model can analyze large volumes of data and identify patterns and insights that are not easily visible to human analysts. Counterfactuals and what-if analysis can be used to test and validate the accuracy of the machine learning models and get ideas on retaining the customers who are likely to churn, leading to higher customer retention.

- **Personalized solutions:** Machine learning can analyze customer behavior and preferences to offer personalized solutions that can help retain customers. Counterfactuals and what-if analysis can be used to test different scenarios and predict how customers will respond to different solutions, leading to more effective personalized solutions.

- **Faster decision-making:** Machine learning can process data in real time and offer insights that can help decision-makers make faster and more informed decisions. Counterfactuals and what-if analysis can be used to test different scenarios and predict how different decisions will impact customer churn, enabling decision-makers to make more informed decisions more quickly.

- **Cost-effective:** The proposed solution can automate many tasks that would otherwise require human analysts, leading to cost savings for the company. By automating routine tasks, such as data analysis and report generation, human analysts can focus on more complex tasks that require human intuition and expertise.

- **Scalability:** The solution can handle large volumes of data and scale as the customer base grows. This can be particularly useful for a company that is experiencing rapid growth and needs a solution that can keep up with its expanding customer base.

- **Continuous improvement:** The solution can learn from new data and continuously improve their predictions and recommendations. This can lead to better outcomes over time as the decision intelligence system becomes more sophisticated and accurate.

- **Revenue growth:** Retaining current customers can contribute to revenue growth, as satisfied customers are more likely to purchase additional products and services.

- **Competitive advantage:** Successfully predicting and mitigating customer churn can set the company apart from its competitors. By delivering exceptional customer experiences and satisfying customer demands, it can establish a devoted customer base and stand out in the marketplace.

Dataset Details:

The dataset on customer churn pertains to a hypothetical telecom business that operated in California during the third quarter and encompasses information on 7,043 customers who had either churned, retained, or acquired the company's services. The dataset further features various crucial demographic variables for each customer.

Demographics:

- **CustomerID:** A unique ID that identifies each customer.

- **Gender:** The customer's gender: Male, Female.

- **Senior Citizen:** Indicates if the customer is 65 or older: Yes, No.

- **Partner:** Indicates if the customer has a partner: Yes, No.

CHAPTER 7   CASE STUDIES

- **Dependents:** Indicates if the customer lives with any dependents: Yes, No. Dependents could be children, parents, grandparents, etc.

Services:

- **Tenure in Months:** Indicates the total number of months that the customer has been with the company by the end of the quarter.

- **Phone Service:** Indicates if the customer subscribes to home phone service with the company: Yes, No.

- **Multiple Lines:** Indicates if the customer subscribes to multiple telephone lines with the company: Yes, No.

- **Internet Service:** Indicates if the customer subscribes to Internet service with the company: No, DSL, Fiber Optic, Cable.

- **Online Security:** Indicates if the customer subscribes to an additional online security service provided by the company: Yes, No.

- **Online Backup:** Indicates if the customer subscribes to an additional online backup service provided by the company: Yes, No.

- **Device Protection Plan:** Indicates if the customer subscribes to an additional device protection plan for their Internet equipment provided by the company: Yes, No.

- **Premium Tech Support:** Indicates if the customer subscribes to an additional technical support plan from the company with reduced wait times: Yes, No.

- **Streaming TV:** Indicates if the customer uses their Internet service to stream television programming from a third party provider: Yes, No. The company does not charge an additional fee for this service.

- **Streaming Movies:** Indicates if the customer uses their Internet service to stream movies from a third party provider: Yes, No. The company does not charge an additional fee for this service.

- **Contract:** Indicates the customer's current contract type: Month-to-Month, One Year, Two Year.

- **Paperless Billing:** Indicates if the customer has chosen paperless billing: Yes, No.

- **Payment Method:** Indicates how the customer pays their bill: Bank Withdrawal, Credit Card, Mailed Check.

- **Monthly Charge:** Indicates the customer's current total monthly charge for all their services from the company.

- **Total Charges:** Indicates the customer's total charges, calculated to the end of the quarter.

Churn Status:

- **Churn Label:** Yes = the customer left the company this quarter. No = the customer remained with the company. This is directly related to the churn value.

Source: `https://community.ibm.com/community/user/ businessanalytics/blogs/steven-macko/2019/07/11/telco-customer- churn-1113`

Stage 1: AI Model Creation

The goal of this activity is to build a machine learning model that can fairly predict the customer churn. Although advanced data processing techniques and ML algorithms are available, we will use the ones that give satisfactory results to ensure we are spending more efforts toward the goal, i.e., building a decision intelligence system.

Step 1: Importing the required libraries

```
[In]:
import dice_ml
import numpy as np
import pandas as pd
from sklearn.preprocessing import LabelEncoder
from sklearn.ensemble import RandomForestClassifier
from sklearn.model_selection import RandomizedSearchCV,
train_test_split
from sklearn.metrics import accuracy_score, precision_score,
recall_score, f1_score
```

Step 2: Loading the telecom churn dataset

```
[In]:
churn_data_all = pd.read_csv('telco_churn.csv')
```

Step 3: Checking the data characteristics to see whether there are any discrepancies in the data (e.g., missing values, wrong data types, etc.)

```
[In]:
churn_data_all.info()
[Out]:
<class 'pandas.core.frame.DataFrame'>
RangeIndex: 7043 entries, 0 to 7042
Data columns (total 21 columns):
 #   Column            Non-Null Count   Dtype
---  ------            --------------   -----
 0   customerID        7043 non-null    object
```

```
1    gender            7043 non-null    object
2    SeniorCitizen     7043 non-null    int64
3    Partner           7043 non-null    object
4    Dependents        7043 non-null    object
5    tenure            7043 non-null    int64
6    PhoneService      7043 non-null    object
7    MultipleLines     7043 non-null    object
8    InternetService   7043 non-null    object
9    OnlineSecurity    7043 non-null    object
10   OnlineBackup      7043 non-null    object
11   DeviceProtection  7043 non-null    object
12   TechSupport       7043 non-null    object
13   StreamingTV       7043 non-null    object
14   StreamingMovies   7043 non-null    object
15   Contract          7043 non-null    object
16   PaperlessBilling  7043 non-null    object
17   PaymentMethod     7043 non-null    object
18   MonthlyCharges    7043 non-null    float64
19   TotalCharges      7032 non-null    float64
20   Churn             7043 non-null    object
dtypes: float64(2), int64(2), object(17)
memory usage: 1.1+ MB
```

Step 4: Data preprocessing

All columns except the TotalCharges do not have any null values. The TotalCharges column needs to be treated with some appropriate missing value strategy. Let's look at the rows where the column has null values.

```
[In]:
churn_data_all[churn_data_all['TotalCharges'].isnull()].head()
[Out]:
```

customerID	gender	SeniorCitizen	Partner	Dependents	tenure	Phone Service	MultipleLines	Total Charges	Churn
4472-LVYGI	Female	0	Yes	Yes	0	No	No phone service		No
3115-CZMZD	Male	0	No	Yes	0	Yes	No		No
5709-LVOEQ	Female	0	Yes	Yes	0	Yes	No		No
4367-NUYAO	Male	0	Yes	Yes	0	Yes	Yes		No
1371-DWPAZ	Female	0	Yes	Yes	0	No	No phone service		No
7644-OMVMY	Male	0	Yes	Yes	0	Yes	No		No
3213-VVOLG	Male	0	Yes	Yes	0	Yes	Yes		No
2520-SGTTA	Female	0	Yes	Yes	0	Yes	No		No
2923-ARZLG	Male	0	Yes	Yes	0	Yes	No		No
4075-WKNIU	Female	0	Yes	Yes	0	Yes	Yes		No
2775-SEFEE	Male	0	No	Yes	0	Yes	Yes		No

(Some columns are not shown in the previous table to fit within the viewable window.)

From the previous table, we find that the TotalCharges column is null when the tenure is 0. This means the customers who have joined recently and haven't even completed 1 month with the telecom operator will have their TotalCharges as null. Let's convert the null value to 0, as that would be the most appropriate value.

```
[In]:
churn_data_all['TotalCharges'] = np.where(churn_
data_all['TotalCharges'].isnull(), 0, churn_data_
all['TotalCharges'])
```

While the SeniorCitizen column contains binary values, its encoding differs from that of other attributes. Therefore, we convert it to a format that conforms to a standard representation for all variables.

```
[In]:
churn_data_all['SeniorCitizen'] = np.where(churn_data_all.
SeniorCitizen == 1,"Yes","No")
```

We create a list of column names based on their type and use that.

```
[In]:
all_columns = [x for x in churn_data_all.drop('customerID',
axis = 1).columns]
id_column = ['customerID']
target_column = ['Churn']
categorical_columns = [y for y in churn_data_all.
drop('customerID', axis = 1).select_dtypes(include = [object]).
columns]
numeric_columns = [z for z in all_columns if z not in
categorical_columns]
[In]:
```

```
get_dummies = []
label_encoding = []
for i in categorical_columns:
    print('Column Name:', i, ', Unique Value Counts:',
    len(churn_data_all[i].unique()), ', Values:', churn_data_
    all[i].unique())
    if len(churn_data_all[i].unique()) > 2:
        get_dummies.append(i)
    else:
        label_encoding.append(i)
```

[Out]:

Column Name: gender , Unique Value Counts: 2 , Values:
['Female' 'Male']

Column Name: SeniorCitizen , Unique Value Counts: 2 , Values:
['No' 'Yes']

Column Name: Partner , Unique Value Counts: 2 , Values:
['Yes' 'No']

Column Name: Dependents , Unique Value Counts: 2 , Values:
['No' 'Yes']

Column Name: PhoneService , Unique Value Counts: 2 , Values:
['No' 'Yes']

Column Name: MultipleLines , Unique Value Counts: 3 , Values:
['No phone service' 'No' 'Yes']

Column Name: InternetService , Unique Value Counts: 3 , Values:
['DSL' 'Fiber optic' 'No']

Column Name: OnlineSecurity , Unique Value Counts: 3 , Values:
['No' 'Yes' 'No internet service']

Column Name: OnlineBackup , Unique Value Counts: 3 , Values:
['Yes' 'No' 'No internet service']

Column Name: DeviceProtection , Unique Value Counts: 3 ,
Values: ['No' 'Yes' 'No internet service']

Column Name: TechSupport , Unique Value Counts: 3 , Values:
['No' 'Yes' 'No internet service']
Column Name: StreamingTV , Unique Value Counts: 3 , Values:
['No' 'Yes' 'No internet service']
Column Name: StreamingMovies , Unique Value Counts: 3 , Values:
['No' 'Yes' 'No internet service']
Column Name: Contract , Unique Value Counts: 3 , Values:
['Month-to-month' 'One year' 'Two year']
Column Name: PaperlessBilling , Unique Value Counts: 2 ,
Values: ['Yes' 'No']
Column Name: PaymentMethod , Unique Value Counts: 4 , Values:
['Electronic check' 'Mailed check' 'Bank transfer (automatic)'
 'Credit card (automatic)']
Column Name: Churn , Unique Value Counts: 2 , Values:
['No' 'Yes']

We can see that some categorical columns have two unique values, whereas some have more than two. To apply appropriate techniques, we have split the columns.

We apply dummy variable creation techniques to columns having more than two unique values.

```
[In]:
churn_data_all_dl = pd.get_dummies(churn_data_all, prefix=get_
dummies, columns=get_dummies)
```

We apply a label encoding technique to the columns with two unique values and save the mappings.

```
mappings = {}
for col in label_encoding:
    le = LabelEncoder()
```

```
churn_data_all_dl[col] = le.fit_transform(churn_data_all_
dl[col])
mappings[col] = dict(zip(le.classes_,range(len(le.
classes_))))
mappings
[Out]:
{'gender': {'Female': 0, 'Male': 1},
 'SeniorCitizen': {'No': 0, 'Yes': 1},
 'Partner': {'No': 0, 'Yes': 1},
 'Dependents': {'No': 0, 'Yes': 1},
 'PhoneService': {'No': 0, 'Yes': 1},
 'PaperlessBilling': {'No': 0, 'Yes': 1},
 'Churn': {'No': 0, 'Yes': 1}}
```

Step 5: Modeling

We split the dataset for training and inference.

```
[In]:
X = churn_data_all_dl.drop(['customerID', 'Churn'], axis=1)
y = churn_data_all_dl['Churn']
X_train, X_test, y_train, y_test = train_test_split(X, y, test_
size = 0.2, stratify=y, random_state = 0)
```

Hyperparameter Tuning

The process of hyperparameter tuning entails identifying the best hyperparameter combination for a learning algorithm, which can be employed to optimize its performance on any given dataset. By minimizing a prespecified loss function, the selected hyperparameters can reduce errors and improve the model's results.

Running the following lines of code aids in determining the ideal hyperparameter for our machine learning algorithm:

```
[In]:
n_estimators = [int(x) for x in np.linspace(start = 100,
stop = 2100, num = 6)]
feature_name = list(X_test.columns)
max_depth = [int(x) for x in np.linspace(10, 100, num = 5)]
max_depth.append(None)
min_samples_split = [2, 5, 10]
min_samples_leaf = [1, 2, 4, 6, 8, 10]
random_grid = {'n_estimators':n_estimators,
               'max_depth': max_depth,
               'min_samples_split': min_samples_split,
               'min_samples_leaf': min_samples_leaf}
print(random_grid)
rf = RandomForestClassifier()
rf_random = RandomizedSearchCV(estimator = rf, param_
distributions = random_grid, n_iter = 10, cv = 2, verbose=2,
random_state=42, n_jobs = -1)
rf_random.fit(X_train[feature_name], y_train)
print(rf_random.best_params_)
[Out]:
{'n_estimators': [100, 500, 900, 1300, 1700, 2100], 'max_
depth': [10, 32, 55, 77, 100, None], 'min_samples_split':
[2, 5, 10], 'min_samples_leaf': [1, 2, 4, 6, 8, 10]}
```

We fit 2 folds for each of 10 candidates, totaling 20 fits.

```
{'n_estimators': 100, 'min_samples_split': 2, 'min_samples_
leaf': 8, 'max_depth': 10}
```

We see that for the random forest model (the algorithm of choice in this case), the best combination of hyperparameters is as follows:

- n_estimators = 100

- min_samples_split = 2

- min_samples_leaf = 8

- max_depth = 10

We use the previous hyperparameters to build the random forest model.

```
[In]:
feature_name = list(X_test.columns)
churn_classifier=RandomForestClassifier(n_estimators=100,min_
samples_split=2,min_samples_leaf=8,max_depth=10)
churn_classifier.fit(X_train[feature_name],y_train)
[Out]:
RandomForestClassifier(bootstrap=True, ccp_alpha=0.0, class_
                       weight=None,
                       criterion='gini', max_depth=10, max_
                       features='auto',
                       max_leaf_nodes=None, max_samples=None,
                       min_impurity_decrease=0.0, min_impurity_
                       split=None,
                       min_samples_leaf=8, min_samples_split=2,
                       min_weight_fraction_leaf=0.0,
                       n_estimators=100,
                       n_jobs=None, oob_score=False, random_
                       state=None,
                       verbose=0, warm_start=False)
```

Now, we use the trained model for making predictions on the test data and save them along with the prediction probabilities.

```
[In]:
pred_df = X_test.copy()
pred_df['Churn'] = y_test
pred_df['pred'] = churn_classifier.predict
(X_test[feature_name])
prediction_of_probability = churn_classifier.predict_proba
(X_test[feature_name])
pred_df['prob_0'] = prediction_of_probability[:,0]
pred_df['prob_1'] = prediction_of_probability[:,1]
```

We evaluate the model's performance.

```
[In]:
print("Accuracy: ", (accuracy_score(pred_df['Churn'],
pred_df['pred']))*100)
print("Precision: ", (precision_score(pred_df['Churn'],
pred_df['pred']))*100)
print("Recall: ", (recall_score(pred_df['Churn'], pred_
df['pred']))*100)
print("F1 Score: ", (f1_score(pred_df['Churn'], pred_
df['pred']))*100)
[Out]:
Accuracy:   80.48261178140525
Precision:  67.1280276816609
Recall:   51.87165775401069
F1 Score:  58.521870286576174
```

As mentioned, we are aiming for satisfactory model performance for the demonstration. There are multiple ways to get better model performance; these are some of the prominent ones:

- **Feature engineering:** This involves creating new features or transforming existing features to make them more informative for the model.

- **Regularization:** This involves adding a penalty term to the loss function to prevent overfitting of the model to the training data.

- **Ensemble methods:** This involves combining multiple models to create a more accurate prediction.

- **Data augmentation:** This involves increasing the size of the training dataset by generating new examples using techniques such as rotating, flipping, or adding noise to the data.

- **Oversampling/undersampling:** This involves balancing imbalanced datasets by either increasing the number of instances in the minority class (oversampling) or decreasing the number of instances in the majority class (undersampling).

- **Transfer learning:** This involves leveraging a pretrained model to solve a similar problem or using a pretrained model as a feature extractor.

- **Gradient clipping:** This involves clipping the gradients during training to prevent them from becoming too large or too small.

- **Early stopping:** This involves stopping the training process when the model's performance on the validation set stops improving.

Now that we have the model and predictions ready, let's move on and start using them for decision-making.

Stage 2: Generating Counterfactuals

We will be using a combination of counterfactual and what-if analysis to drive decision intelligence. Let's look into what they mean and how they can be used.

Counterfactual Analysis:

Counterfactuals refer to a type of analysis that involves identifying the minimal set of changes required to an input instance such that the machine learning model output changes to a desired outcome. In simpler terms, counterfactuals are hypothetical examples of inputs that would lead to a different output from a machine learning model.

Counterfactual analysis can be used to explain the behavior of a machine learning model and to provide insights into how the model can be improved as well as what actions can be taken so that favorable outcomes can be generated. It can also be used for various applications such as fairness analysis, where the minimal set of changes required to achieve a desired outcome can be used to identify potential sources of bias in the model. Additionally, counterfactual analysis can be used for causal inference, where the minimal set of changes can be used to estimate the causal effect of a treatment on an outcome.

We will be using the DiCEML Python package for generating counterfactuals.

DiCELML

DiCEML (Diverse Counterfactual Explanations through Mixed Integer Linear Programming) is a framework that enables the generation of counterfactual examples for machine learning models. DiCEML employs mixed-integer linear programming to find a minimal set of changes to input instances required to change the model output to a desired outcome. It is a state-of-the-art method for generating diverse counterfactual explanations that take into account the constraints and features of the input data and can be used for various applications such as explainable AI, fairness analysis, and causal inference.

Step 6: Building a counterfactual explainer object

```
[In]:
def initialize_counterfactuals(
```

```python
    train_df, model, feature_list, continuous_features, target,
    model_type
):
    """
    Initialize Counterfactual explainer object for the given
    input model and also the feature range dictionary
    (both are used in calculating Counterfactuals in Local
    Explainability)

    Args:
        train_df (dataframe) : train dataframe
        model (object) : input model (to explain)
        feature_list (list) : list of features used in model
        continuous_features (list) : list of continuous
        features
        target (str) : target column name
        model_type (str) : classification or regression

    Returns:
        explainer (object) : counterfactual explainer object
        feature_range (dict) : dictionary containing
        permissible feature ranges for continuous features
    """
    df_model = train_df[feature_list + [target]]

    ####Round the decimal upto 4 digits
    df_model[continuous_features] = (
        df_model[continuous_features].astype(float).round(4)
    )  # .astype(str)
    print("continuous features", continuous_features)

    #### dice data  initialisations
    data_dice = dice_ml.Data(
```

```
        dataframe=df_model,  # For perturbation strategy
        continuous_features=continuous_features,
        outcome_name=target,
    )

    ## dice model initialisation
    if model_type == "regression" or model_type == "time
    series":
        model_dice = dice_ml.Model(model=model,
        backend="sklearn", model_type="regressor")
    else:
        model_dice = dice_ml.Model(model=model,
        backend="sklearn")

    # getting model and data together
    explainer = dice_ml.Dice(data_dice, model_dice,
    method="random")
    df_model.drop(target, axis=1, inplace=True)
    feature_range = {}

    for i in continuous_features:
        feature_range[i] = [
            df_model[i].astype(float).min(),
            df_model[i].astype(float).max(),
        ]

    return explainer, feature_range

def preprocess_encode(df,mappings,feature_list):
    if mappings != None:
        encode_df = df[feature_list]
        cat_cols = (encode_df.dtypes == object)
        cat_cols = list(cat_cols[cat_cols==True].index)
```

```
        encode_df = df.copy()
        if len(cat_cols)>0:
            encode_df.update(encode_df[cat_cols].apply(lambda
            col: col.map(mappings[col.name])).astype(int))
            encode_df[cat_cols] = encode_df[cat_cols].
            astype(int)
    else:
        encode_df = df.copy()
    return encode_df

def postprocess_decode(df,mappings):
    decoded_output = df.copy()
    if mappings != None:
        inv_mapping_dict = {cat: {v: k for k, v in map_dict.
        items()} for cat, map_dict in mappings.items()}
        decoded_output[list(mappings.keys())] = decoded_
        output[list(mappings.keys())].astype(int)
        decoded_output = decoded_output.replace(inv_
        mapping_dict)
    return decoded_output

def generate_output_counterfactuals(explainer,feature_
range,query_instance,model_type,model,feature_list,desired_
class_or_range,continuous_features,features_vary='all',num_
cf=3,mappings={}):
    """
```

Generating the counterfactuals output for a particular
query_instance (record), given various other inputs like
features to vary, etc.

Args:
    explainer (object) : Counterfactuals explainer object

```
    feature_range (dict) : dictionary containing
    permissible feature ranges for continuous features
    query_instance (dataframe) : selected record to explain
    (user input)
    model_type (str) : classification or regression
    model (object) : model object, required only in the
    case of classification
    feature_list (list) : list of features used in model
    desired_class_or_range (list) : if classification then
    desired output class; if regression then desired output
    range (user input)
    continuous_features (list) : list of continuous
    features
    features_vary (list) : list of features that can be
    varied (user input)
    num_cf (int) : number of counterfactuals to generate
    mappings (dict) : encoder mapping dictionary
Returns:
    cf_output (dataframe) : counterfactuals output
    dataframe
"""
query_instance[continuous_features] = query_
 instance[continuous_features].astype(float).round(4)
query_instance_processed = preprocess_encode(query_
instance,mappings,list(query_instance.columns))
display(query_instance_processed)
if model_type == 'classification':
    cf_exp = explainer.generate_counterfactuals(query_
    instance_processed, total_CFs=20,desired_class=desired_
    class_or_range,permitted_range=feature_range,features_
    to_vary=features_vary)
```

```
    else:
        cf_exp = explainer.generate_counterfactuals(query_
        instance_processed, total_CFs=num_cf,desired_
        range=desired_class_or_range,permitted_range=feature_
        range,features_to_vary=features_vary)
    cf_exp_df = cf_exp.cf_examples_list[0].final_cfs_df

    if model_type == "classification":
        cf_exp_df["Probability"] = model.predict_proba(cf_exp_
        df[feature_list])[:, desired_class_or_range]
        cf_exp_df = (cf_exp_df.sort_values("Probability",
        ascending=False).reset_index(drop=True).head(num_cf))

    cf_exp_df = postprocess_decode(cf_exp_df,mappings)
    cf_output = cf_ouput_df_fn(cf_exp_df,query_instance)
    return cf_output

def cf_ouput_df_fn(df,org_df):
    newdf = df.values.tolist()
    org = org_df.values.tolist()[0]
    for ix in range(df.shape[0]):
        for jx in range(len(org)):
            if str(newdf[ix][jx]) == str(org[jx]):
                newdf[ix][jx] = '-'
            else:
                newdf[ix][jx] = str(newdf[ix][jx])
    return pd.DataFrame(newdf, columns=df.columns,
    index=df.index)

features = [x for x in X_train.columns]

train_df = X_train.copy()
train_df["Churn"] = y_train
```

```
model_type='classification'
target = 'Churn'
explainer, feature_range = initialize_counterfactuals(
    train_df, churn_classifier, features, numeric_columns,
    target, model_type
)
[Out]:
continuous features ['tenure', 'MonthlyCharges',
'TotalCharges']
```

Stage 3: What-If Analysis

Step 7: Using the explainer object to generate possible strategies to retain customers

Now we have the explainer counterfactual object that can be used to produce counterfactual explanations, i.e., what changes in the inputs can be made to get the desired output. In our case, it would be: "For the customers who are likely to churn, what can be changed (services, account related, demographics) so that he/she is retained." Let's understand this by looking at an example.

The following are the inputs DiCEML expects:

- The trained machine learning model.

- The input instance for which we want to generate a counterfactual. We will take a sample user from a pool of users who are predicted to be churning.

- The features to vary. For this example, we will see if providing services such as paperless billing, online security, online backup, device protection, tech support, Internet services for streaming TV or movies, one- or two-year contract, or if changing the monthly plan or a combination of features can change the customer's mind and they might plan to stay with the telecom operator.

- The desired output or the target class for which we
  want to generate a counterfactual. In this case, it will be
  nonchurn, i.e., 0 (from 1).

Here's how it works:

```
[In]:
test_df = X_test.copy()
test_df["Churn"] = y_test

test_df_churners = pred_df[pred_df['Churn'] == 1].sort_
values(by=['prob_1'], ascending=False)

test_df_churners_input = test_df_churners.drop(['Churn',
'pred', 'prob_0', 'prob_1'], inplace=False, axis=1)

test_df_churners_input_record = test_df_churners_input[61:62]

features_vary = ['PaperlessBilling', 'MonthlyCharges',
                 'OnlineSecurity_Yes', 'OnlineBackup_Yes',
                 'DeviceProtection_Yes', 'TechSupport_Yes',
                 'StreamingTV_Yes', 'StreamingMovies_Yes',
                 'Contract_One year', 'Contract_Two year']

dice_exp = explainer.generate_counterfactuals(test_df_churners_
input_record, total_CFs=4, desired_class="opposite", features_
to_vary=features_vary)
dice_exp.visualize_as_dataframe()
[Out]:
100%|████████████████████████| 1/1 [00:31<00:00, 31.02s/it]
[In]:
test_df_churners_input_record[features_vary]
[Out]:
```

Paperless Billing	Monthly Charges	Online Security_ Yes	Online Backup_Yes	Device Protection_Yes	Tech Support_Yes	Streaming TV_Yes	Streaming Movies_ Yes	Contract_ One year	Contract_ Two year
0	101.15	0	0	1	0	1	1	0	0

```
[In]:
dice_exp.cf_examples_list[0].final_cfs_df[features_vary]
```

Paperless Billing	Monthly Charges	Online Security_Yes	Online Backup_Yes	Device Protection_ Yes	Tech Support_ Yes	Streaming TV_Yes	Streaming Movies_ Yes	Contract_ One year	Contract_ Two year
0	55.36	1	0	1	0	1	1	0	0
0	35.86	1	0	1	0	1	1	0	0
0	22.32	1	0	1	0	1	1	0	0
0	67.89	1	0	1	0	1	1	0	0

From the previous output, we see that the explainer model has suggested four possible ways to retain the customer.

- Change monthly charges from 101.15 to 55.36 and provide online security service.

- Change monthly charges from 101.15 to 35.86 and provide online security service.

- Change monthly charges from 101.15 to 22.32 and provide online security service.

- Change monthly charges from 101.15 to 67.89 and provide online security service.

However, there might be certain constraints for the business to take into consideration, and one cannot rely only on the suggestions from the counterfactual analysis. One might want to take the inputs from the counterfactual analysis, tweak them a bit based on the business constraints, and then check if those tweaked inputs generate the desired output. This is possible through what-if analysis.

Step 8: Using what-if analysis for getting the right set of strategies to retain customers (based on constraints)

What-if analysis is a technique used to understand the behavior of a machine learning model under different scenarios, or what-if, situations. It involves testing the model with different input values or scenarios to understand how the model behaves in those situations.

Now suppose for the given counterfactual recommendations, we have a constraint that the customer monthly charges cannot be reduced by more than 10 percent, as it can result in large revenue losses. So, for the given customer, we cannot bring the monthly charges below 91. Also, we do not have any constraint on providing services for free, so let's see with the given constrained monthly charges and a combination of services if we can retain the customer or not.

Options:

- Change monthly charges from 101.7 to 91; provide online security and online backup services.

```
[In]:
custom_input = test_df_churners_input_record.copy()
custom_input['MonthlyCharges'] = custom_
input['MonthlyCharges'].replace(101.15, 91)
custom_input['OnlineBackup_Yes'] = custom_
input['OnlineBackup_Yes'].replace(0, 1)
custom_input['OnlineSecurity_Yes'] = custom_
input['OnlineSecurity_Yes'].replace(0, 1)

if churn_classifier.predict(custom_input)==0:
    print("Successfully retained the customer")
else:
    print("Sorry, customer could not be retained")

print("\nProbability of Churn:",churn_classifier.
predict_proba(custom_input)[0][1].round(2))

[Out]:
Sorry, customer could not be retained
Probability of Churn: 0.57
```

We see that with the given inputs, we cannot retain the customer. Let's try another set of inputs,

- Change monthly charges from 101.7 to 91; provide online security and online backup services with a 2-year contract.

```
[In]:
custom_input2 = test_df_churners_input_record.copy()
```

```
custom_input2['MonthlyCharges'] = custom_
input2['MonthlyCharges'].replace(101.15, 91)
custom_input2['OnlineBackup_Yes'] = custom_
input2['OnlineBackup_Yes'].replace(0, 1)
custom_input2['OnlineSecurity_Yes'] = custom_
input2['OnlineSecurity_Yes'].replace(0, 1)
custom_input2['Contract_Two year'] = custom_
input2['Contract_Two year'].replace(0, 1)

if churn_classifier.predict(custom_input2)==0:
    print("Successfully retained the customer")
else:
    print("Sorry, customer could not be retained")

print("\nProbability of Churn:",churn_classifier.
predict_proba(custom_input2)[0][1].round(2))
[Out]:
Successfully retained the customer
Probability of Churn: 0.47
```

We can see that with the given set of inputs in this option that we were successful in retaining the customer with a churn probability of the customer coming down to 47 percent. So, this option can be converted to a personalized offering to the customer.

Counterfactuals and "what-if" analysis can be used to detect bias in machine learning models. It involves analyzing counterfactual outputs or testing the model's sensitivity to changes in the input data or model parameters to identify areas where the model may be biased. Let's see how bias can be detected using counterfactual analysis + what-if for the same previously used record, adding demographics SeniorCitizen and gender into the features to vary the matrix.

```
[In]:
test_df_churners_bias_inputs = test_df_churners_input[(test_
df_churners_input['SeniorCitizen'] == 1) | (test_df_churners_
input['gender'] == 1)]
features_vary_bias = ['SeniorCitizen', 'gender']
dice_exp_bias = explainer.generate_counterfactuals(test_
df_churners_bias_inputs[100:130], total_CFs=4, desired_
class="opposite", features_to_vary=features_vary_bias)
dice_exp_bias.visualize_as_dataframe()
```

The previous counterfactual output, in a few instances, provides recommendations to change either the gender or age (senior citizen to nonsenior citizen) for favorable outcomes. This indicates there is a need to check for bias in the model. It is important to note that these methods are not foolproof for detecting and mitigating bias in AI systems, and it is crucial to use a combination of techniques, data validation, auditing, and diversity and inclusion practices to ensure AI systems produce equitable outcomes. Fairness metrics such as disparate impact, equal opportunity difference, false positive rate parity, equalized odds, and calibration are used to assess whether an AI model is treating different groups of individuals fairly, and the choice of metrics depends on the specific context and goals. Disparate impact is one such measure of fairness that examines whether the AI model's outcomes have a disproportionate impact on particular groups based on protected characteristics, and its formula is used to calculate the disparate impact ratio for two groups being compared. Let's calculate the disparate impact value for the Gender and SeniorCitizen columns in our dataset.

```
[In]:
protected_columns = ["gender", "SeniorCitizen"]
target="pred"
for pro_col in protected_columns:
    categories = list(pred_df[pro_col].unique())
```

```
selection_rate_list = []
for cat in categories:
    selection_rate = pred_df[(pred_df[target]==1)&(pred_
    df[pro_col]==cat)].shape[0]/pred_df[pred_df[pro_
    col]==cat].shape[0]
    selection_rate_list.append(np.round(selection_rate,3))
disparate_impact = np.round(min(selection_rate_list)/
max(selection_rate_list),3)
print("Disparate Impact for",pro_col,":",disparate_impact)
[Out]:
Disparate Impact for gender : 0.843
Disparate Impact for SeniorCitizen : 0.438
```

The disparate impact score for the SeniorCitizen column is quite low, suggesting a bias toward this feature, while the Gender column has a score of about 84%, indicating a very fair treatment, which exceeds the threshold of 80 percent. However, the SeniorCitizen score of about 44 percent is significantly lower than the threshold value, indicating an unfair bias. It is important to use multiple fairness metrics in conjunction with each other to arrive at a conclusion.

Case Study 2: Mobile Phone Pricing/ Configuration Strategy

Problem Statement: A mobile phone manufacturing company wants to improve its pricing strategy by utilizing the power of AI and machine learning algorithms. The company has a large database of mobile phone models with various features such as screen size, camera quality, processor speed, battery life, and others. However, it currently lacks an accurate and efficient method for predicting the prices of its mobile phones based on these features. The company is seeking a solution that can help them forecast prices for new phone models with a high degree of accuracy, allowing them to optimize their pricing strategy and stay competitive in the market.

Solution:

The AI team proposes a hybrid solution to solve the given problem. The plan is to first build a machine learning model that takes historical mobile details such as screen size, camera quality, processor speed, battery life, and others along with the price. The model output will be used to predict the potential price of the handset. The team also proposes a counterfactual generation engine that can provide various scenarios wherein the predicted outcomes change as per the expected price range. For example, if the price of a handset is predicted to be X and if the manufacturer wants to sell it at a price 10 percent to 15 percent more than the predicted one, the engine will suggest changes in the phone's configuration to the expected price range. This change in the input parameters can be used as part of product strategy. A third component of a what-if analysis solution can help the decision-makers to tweak the counterfactual suggestions based on certain constraints and check whether those changes can still help them sell the phone at the expected price range.

The AI team suggests the following benefits of using the proposed hybrid solution for mobile pricing prediction:

- **Accurate price predictions:** AI algorithms can analyze vast amounts of data and identify patterns that humans may miss. This enables the tool to accurately predict mobile handset prices based on various features with greater accuracy.

- **Real-time pricing:** With an AI-driven tool, pricing decisions can be made in real time based on changes in market demand and supply. This allows the mobile phone manufacturer to respond quickly to changing market conditions and adjust its pricing strategies accordingly.

- **Improved profit margins:** By accurately predicting prices, the manufacturer can optimize its pricing strategies to maximize profit margins. The tool can identify the optimal price point for a product, taking into account factors such as production costs, competition, and market demand.

- **Increased customer satisfaction:** By setting the right price for a product, the manufacturer can increase customer satisfaction. Customers are more likely to purchase a product if they feel that it is priced fairly and accurately reflects its value.

- **Better decision-making:** An AI-driven tool can provide the manufacturer with valuable insights into customer preferences and buying behavior. This can help make informed decisions about product development, marketing, and pricing strategies.

Dataset Details:

These are the dataset details:

- **Product_id:** ID of each cell phone

- **Price:** Price of each cell phone

- **Sale:** Sales number

- **weight:** Weight of each cell phone

- **resolution:** Resolution of each cell phone

- **ppi:** Phone pixel density

- **cpu core:** Type of CPU core in each cell phone

175

- **cpu freq:** CPU frequency in each cell phone

- **internal mem:** Internal memory of each cell phone

- **ram:** RAM of each cell phone

- **RearCam:** Resolution of rear camera

- **Front_Cam:** Resolution of front camera

- **battery:** Battery capacity (in mA)

- **thickness:** Thickness of the phone

Source: https://www.kaggle.com/datasets/mohannapd/mobile-price-prediction

Stage 1: AI Model Creation

The goal of this activity is to build a machine learning model that can fairly predict mobile prices. Although advanced data processing techniques and ML algorithms are available, we will use the ones that give satisfactory results to ensure we are spending more efforts toward the goal, i.e., building a decision intelligence system.

Step 1: Importing the required libraries

```
[In]:
import pandas as pd
import dice_ml
from dice_ml import Dice
from sklearn.metrics import r2_score
from sklearn.ensemble import RandomForestRegressor
from sklearn.model_selection import train_test_split
```

Step 2: Loading the dataset

```
[In]:
all_data = pd.read_csv('cellphone_price_data.csv')
all_data.head()

[Out]:
```

Product_id	Price	Sale	weight	resolution	ppi	cpu core	cpu freq	internal mem	ram	RearCam	Front_Cam	battery	thickness
203	2357	10	135	5.2	424	8	1.35	16	3	13	8	2610	7.4
880	1749	10	125	4	233	2	1.3	4	1	3.15	0	1700	9.9
40	1916	10	110	4.7	312	4	1.2	8	1.5	13	5	2000	7.6
99	1315	11	118.5	4	233	2	1.3	4	0.512	3.15	0	1400	11
880	1749	11	125	4	233	2	1.3	4	1	3.15	0	1700	9.9

Step 3: Checking the data characteristics to see whether there are any discrepancies in the data (e.g., missing values, wrong data types, etc.)

```
[In]:
all_data.info()
[Out]:
<class 'pandas.core.frame.DataFrame'>
RangeIndex: 161 entries, 0 to 160
Data columns (total 14 columns):
 #   Column        Non-Null Count   Dtype
---  ------        --------------   -----
 0   Product_id    161 non-null     int64
 1   Price         161 non-null     int64
 2   Sale          161 non-null     int64
 3   weight        161 non-null     float64
 4   resolution    161 non-null     float64
 5   ppi           161 non-null     int64
 6   cpu core      161 non-null     int64
 7   cpu freq      161 non-null     float64
 8   internal mem  161 non-null     float64
 9   ram           161 non-null     float64
 10  RearCam       161 non-null     float64
 11  Front_Cam     161 non-null     float64
 12  battery       161 non-null     int64
 13  thickness     161 non-null     float64
dtypes: float64(8), int64(6)
memory usage: 17.7 KB
```

As we see from the previous details, the dataset does not have any missing values or any issues related to the data. So, no data preprocessing is required here, and we can directly move on to the modeling.

Step 4: Modeling

We split the dataset for training and inference.

```
[In]:
x=all_data.drop(['Price', 'Product_id'],axis=1)
y=all_data['Price']
x_train,x_test,y_train,y_test=train_test_split(x,y,test_
size=0.2,random_state=0)
```

Now, we build a random forest regressor model and use it to make predictions on the test data, saving and evaluating them.

```
[In]:
rf = RandomForestRegressor()
rf.fit(x_train,y_train)
[Out]:
RandomForestRegressor()

[In]:
predictions = rf.predict(x_test)
print('R2 score for Training Data: ', rf.score(x_train,
y_train))
print('R2 score for Inference Data: ', r2_score(y_
test,predictions))
[Out]:
R2 score for Training Data:  0.9945696702046988
R2 score for Inference Data:  0.9678002631345382
[In]:
x_test['predicted_price'] = predictions
x_train.info()
[Out]:
<class 'pandas.core.frame.DataFrame'>
Int64Index: 128 entries, 80 to 47
Data columns (total 12 columns):
```

#	Column	Non-Null Count	Dtype
0	Sale	128 non-null	int64
1	weight	128 non-null	float64
2	resolution	128 non-null	float64
3	ppi	128 non-null	int64
4	cpu core	128 non-null	int64
5	cpu freq	128 non-null	float64
6	internal mem	128 non-null	float64
7	ram	128 non-null	float64
8	RearCam	128 non-null	float64
9	Front_Cam	128 non-null	float64
10	battery	128 non-null	int64
11	thickness	128 non-null	float64

```
dtypes: float64(8), int64(4)
memory usage: 13.0 KB
[In]:
x_train['Price'] = all_data['Price']
```

Stage 2: Generating Counterfactuals

We will be using a combination of counterfactual and what-if analysis to drive decision intelligence. Unlike in the classification use case, we will be using the expected price range of the mobile as the desired outcome as this is a regression use case.

```
[In]:
d_mobile = dice_ml.Data(dataframe=x_train, continuous_
features=['Sale', 'weight', 'resolution', 'ppi', 'cpu core',
'cpu freq', 'internal mem', 'ram', 'RearCam', 'Front_Cam',
'battery', 'thickness'], outcome_name='Price')
m_mobile = dice_ml.Model(model=rf, backend="sklearn", model_
type='regressor')

explainer_mobile = Dice(d_mobile, m_mobile, method="random")
```

Stage 3: What-If Analysis

Step 5: Using the explainer object to generate possible pricing strategies

Now we have the explainer counterfactual object that can be used to produce counterfactual explanations, i.e., what changes in the inputs can be made to get the desired output. In our case, it would be: "For the given predicted price, what can be changed (mobile configurations like screen size, resolution, etc.) so that the mobile can be sold at a higher price." Let's understand this through an example.

```
[In]:
input_data = x_test.drop(['predicted_price'], axis=1)
input_record = input_data[0:1]
input_record.head()
[Out]:
```

Sale	weight	resolution	ppi	cpu core	cpu freq	internal mem	ram	RearCam	Front_Cam	battery	thickness
302	149	5.5	534	8	1.6	32	3	16	8	3000	7

```
[In]:
x_test[0:1].head()
[Out]:
```

Sale	weight	resolution	ppi	cpu core	cpu freq	internal mem	ram	RearCam	Front_Cam	battery	thickness	predicted_price
302	149	5.5	534	8	1.6	32	3	16	8	3000	7	2941.57

For the given record, we see that the predicted price is about 2.9K. Suppose if the manufacturer wants to sell the mobile at a price 10 percent to 15 percent higher than the predicted one, the explainer can be used to see recommendations in the mobile configuration that can achieve the objective.

```
[In]:
predicted_price = x_test[0:1].predicted_price
expected_price_range = [round(float(predicted_price*1.10), 0),
round(float(predicted_price*1.15), 0)]
random_mobile = explainer_mobile.generate_
counterfactuals(input_record, total_CFs=2,
    desired_range=expected_price_range)
random_mobile.visualize_as_dataframe(show_only_changes=True)
[Out]:
100%|█████████████████████| 1/1 [00:00<00:00,  1.40it/s]
Query instance (original outcome : 2942)
```

Sale	weight	resolution	ppi	cpu core	cpu freq	internal mem	ram	RearCam	Front_ Cam	battery	thickness	Price
302	149	5.5	534	8	1.6	32	3	16	8	3000	7	2942

Diverse Counterfactual set (new outcome: [3236.0, 3383.0])

Sale	weight	resolution	ppi	cpu core	cpu freq	internal mem	ram	RearCam	Front_ Cam	battery	thickness	Price
-	-	-	-	-	1.6	84.8	5.7	-	-	-	-	3278.169922
-	-	-	-	-	1.6	-	5.2	-	-	5116	-	3305.25

From the previous output, we see that the explainer model has suggested two possible ways to sell the mobile at the expected price range.

1. Change the internal memory from 32 to 84.8 GB, and increase the RAM from 3 to 5.7 GB.

2. Change the RAM from 3 to 5.2 GB, and increase battery capacity from 3000 to 5116 mA.

However, there might be certain constraints for the business to take into consideration, and one cannot rely only on the suggestions from the counterfactual analysis. One might want to take the inputs from the counterfactual analysis, tweak them a bit based on the business constraints, and then check if those tweaked inputs generate the desired output. This is possible through what-If analysis.

Step 6: Using What-If analysis for getting the right set of pricing strategies for the given handset

Now suppose for the given counterfactual recommendations, we have a constraint that expensive parts such as RAM cannot be increased, but cheaper components such as internal memory can be increased by two to four times the present one, and battery can be increased by a maximum of 50 percent. So, for the given mobile, we cannot increase the RAM. Let's check which configuration can help us get the desired price.

Here are the options:

1. Increase the internal memory by two times and the battery by 50 percent.

```
[In]:
expected_internal_memory = int(input_
record['internal mem']*2)
expected_battery = int(input_
record['battery']*1.5)
custom_inputs = input_record.copy()
```

```
custom_inputs['internal mem'] = custom_
inputs['internal mem'].replace(int(input_
record['internal mem']), expected_
internal_memory)
custom_inputs['battery'] = custom_
inputs['battery'].replace(int(input_
record['battery']), expected_battery)
print('New price based on updated
configuration: ', rf.predict(custom_inputs))
if rf.predict(custom_inputs)>=expected_price_
range[0]:
    print("The given configuration helps in
    getting the minimum expected price")
else:
    print("The given configuration does not
    help in getting the minimum expected price
    of",expected_price_range[0])
[Out]:
New price based on updated
configuration:  [3043.53]
The given configuration does not help in getting
the minimum expected price of 3236.0
```

We see that increasing the internal memory by 2x and battery size by 50% does not help in getting the expected price range. Let us see if we can achieve the objective with a different combination.

2. Increase the internal memory by four times and the battery by 50 percent.

```
[In]:
expected_internal_memory2 = int(input_
record['internal mem']*4)
expected_battery2 = int(input_
record['battery']*1.5)
custom_inputs2 = input_record.copy()
custom_inputs2['internal mem'] = custom_
inputs2['internal mem'].replace(int(input_
record['internal mem']), expected_internal_
memory2)
custom_inputs2['battery'] = custom_
inputs2['battery'].replace(int(input_
record['battery']), expected_battery2)
print('New price based on updated
configuration: ', rf.predict(custom_inputs2))
if rf.predict(custom_inputs2)>=expected_price_
range[0]:
    print("The given configuration helps in
    getting the minimum expected price")
else:
    print("The given configuration does not
    help in getting the minimum expected price
    of",expected_price_range[0])
[Out]:
New price based on updated
configuration:  [3247.54]
The given configuration helps in getting the
minimum expected price of 3236.0
```

We can see that the second option gives us the right configuration to achieve the objective of increasing the handset price by at least 10 percent of the predicted one.

Conclusion

In this chapter, we saw how decision-making can be supported by applying AI/ML and other advanced techniques (counterfactuals and what-if analysis). We saw how organizations not only can predict which customers are going to churn but also can take proactive measures to avoid the churn, with the help of counterfactuals and what-if analysis. We also saw how a combination of machine learning predictions, counterfactuals, and what-if analysis can help mobile manufacturers design their product configurations (screen size, RAM, etc.) and get the desired price over the predicted ones. Through these use cases we covered decision intelligence for classification and regression problems.

Index

© Akshay Kulkarni, Adarsha Shivananda, Avinash Manure 2023
A. Kulkarni et al., *Introduction to Prescriptive AI*,
https://doi.org/10.1007/978-1-4842-9568-7

Printed in the United States
by Baker & Taylor Publisher Services

Printed in the United States
by Baker & Taylor Publisher Services